计算机英语

主 编 吴海萍 王开艳

参 编 王玲霞 唐晓莉 丁 奕 陈腊梅
　　　杨燕珠 吴 澍 邵小兰 毛艳青

北京理工大学出版社
BEIJING INSTITUTE OF TECHNOLOGY PRESS

内 容 简 介

本书共 10 章，内容包含了计算机专业英语简介、计算机硬件、操作系统、应用软件、多媒体、网络安全、电子商务、新技术等。在每章开始前设有知识目标和技能目标，每课配有相应的词汇表、注释、习题、对话和阅读材料、参考译文，可供学生检查学习效果与自测使用。附录部分列出了词汇表、常用缩略词表、阅读译文、习题答案。在内容选择和编排上，本书充分考虑了当前计算机专业英语发展的现状以及院校的实际需求，遵循了由浅入深、循序渐进的原则。选材力求紧跟计算机的发展步伐，做到内容新、知识面广、词汇量大，内容通俗易懂。

本书可作为计算机及相关专业的计算机专业英语教材，也可作为广大科技人员学习计算机专业英语知识或参加有关计算机专业英语考试的参考用书。

版权专有　侵权必究

图书在版编目（CIP）数据

计算机英语/吴海萍，王开艳主编. —北京：北京理工大学出版社，2020.5（2024.8 重印）

ISBN 978-7-5682-8028-0

Ⅰ. ①计⋯　Ⅱ. ①吴⋯ ②王⋯　Ⅲ. ①电子计算机－英语－高等职业教育－教材　Ⅳ. ①TP3

中国版本图书馆 CIP 数据核字（2020）第 001178 号

出版发行 /	北京理工大学出版社有限责任公司
社　　址 /	北京市海淀区中关村南大街 5 号
邮　　编 /	100081
电　　话 /	（010）68914775（总编室）
	（010）82562903（教材售后服务热线）
	（010）68944723（其他图书服务热线）
网　　址 /	http：//www.bitpress.com.cn
经　　销 /	全国各地新华书店
印　　刷 /	北京国马印刷厂
开　　本 /	787 毫米×1092 毫米　1/16
印　　张 /	8.75
字　　数 /	206 千字
版　　次 /	2020 年 5 月第 1 版　2024 年 8 月第 7 次印刷
定　　价 /	29.80 元

责任编辑 / 梁铜华
文案编辑 / 梁铜华
责任校对 / 刘亚男
责任印制 / 施胜娟

图书出现印装质量问题，请拨打售后服务热线，本社负责调换

前　言

本书学习目标如下：
(1) 能在英文环境下操作计算机，包括硬件的安装和维护、软件的使用等。
(2) 能通过上网或其他途径阅读计算机专业英语文章，从而掌握最新的计算机技术。
(3) 能适应实际工作的需要，包括阅读各类英文计算机技术文档，以及与外国技术专家或同事进行简单的技术交流。

本书特色如下：

1. 英语技能

学习计算机英语，不仅需要了解计算机知识并掌握一定的专业术语，而且还要求具有一定的英语技能。本书的对话部分以每单元的主题为中心，通过情景和交际环境来展示材料，同时为会话表达和口语练习提供范例。对话典型、简练，将接受能力与产出能力的培养有机地结合起来，为学生创设在具体的语境中运用语言的机会。

2. 循序渐进

本书主要以"阅读"为主。在每一个单元中，首先设计了阅读前的活动，帮助学生将新旧知识有机地联系起来。其次，精读部分选用文字浅显易懂的文章，再结合阅读材料，做巩固性、扩展性练习。最后，泛读部分选取最新、最实用的计算机信息和材料，提升学生科技文献的阅读能力，使学生逐步掌握计算机专业英语知识。

3. 学以致用

与其他计算机英语教材不同的是，本书不仅重视学生对专业英语的学习，更侧重实际应用能力的提高。因而既选编了学生较为感兴趣的英语阅读内容，还介绍并拓展了与阅读材料相关的计算机背景知识，注释阅读材料中的相关内容，并扩大学生的知识面。这些内容都相对独立，可以自由选择感兴趣的部分进行学习。

4. 原汁原味

本书所有文章都针对单元主题选取国外的计算机教材、计算机杂志以及计算机网站上的内容，这里包括学术性较强的科技文章，同时考虑到学生主要通过互联网来阅读英文技术文章，还收录了一些比较口语化且具有时效性的"网络文章"。

5. 精简概括

本书就每个章节的主题补充了相对应的"关键术语"。而在"单元测试"中，则以图文结合的形式，巩固这些常用专业术语的记忆，并让学生在具体的语境中掌握并运用这些词

汇，从而可以对计算机英语有一个更全面、更清楚的了解。

总之，本书充分考虑了当前计算机专业英语发展的现状以及高等职业院校的实际需求，遵循了由浅入深、循序渐进的原则。选材力求紧跟计算机的发展步伐，做到内容新、知识面广、词汇量大。全书内容通俗易懂、结构新颖、重点突出、图文并茂。我们期望这本书能成为学习计算机专业英语的好教材、教师授课的好资料。

由于时间仓促，且编者水平有限，加之计算机专业英语发展迅速，书中难免有疏漏和不足之处。我们衷心地希望得到大家的批评指正，以使本书进一步完善。

编　者

… 目　　录 …

Unit 1　What Is Computer English? …………………………………………… 1

　　Section 1：Why ……………………………………………………………… 1
　　Section 2：What …………………………………………………………… 3
　　Section 3：How …………………………………………………………… 4

Unit 2　Computer Hardware ………………………………………………… 6

　　Section 1：Dialogue ……………………………………………………… 6
　　　　What Is Computer Hardware? ………………………………………… 7
　　Section 2：Reading ……………………………………………………… 8
　　　　The Components of Computer Hardware ……………………………… 8
　　Section 3：Computer Terms …………………………………………… 10
　　　　Do You Know? ………………………………………………………… 10
　　Section 4：Exercises …………………………………………………… 11
　　Section 5：Furthering Reading ………………………………………… 13
　　　　History of Computer Hardware ……………………………………… 13

Unit 3　Operating System ………………………………………………… 15

　　Section 1：Dialogue …………………………………………………… 16
　　　　What Operating System Do You Use? ………………………………… 16
　　Section 2：Reading ……………………………………………………… 17
　　　　Operating System ……………………………………………………… 17
　　Section 3：Computer Terms …………………………………………… 20
　　　　Do You Know? ………………………………………………………… 20
　　Section 4：Exercises …………………………………………………… 22
　　Section 5：Furthering Reading ………………………………………… 23
　　　　Android ………………………………………………………………… 23

Unit 4　Application Software ……………………………………………… 25

　　Section 1：Dialogue …………………………………………………… 25

— 1 —

　　　　How to Solve an Office Problem? ………………………………………… 26
　　Section 2：Reading ……………………………………………………………… 27
　　　　Application Software …………………………………………………… 27
　　Section 3：Computer Terms …………………………………………………… 30
　　　　Do You Know? …………………………………………………………… 30
　　Section 4：Exercises …………………………………………………………… 31
　　Section 5：Furthering Reading ………………………………………………… 33
　　　　Microsoft Office ………………………………………………………… 33

Unit 5　Multimedia ……………………………………………………………… 37

　　Section 1：Dialogue …………………………………………………………… 38
　　　　The Ultimate Movie Experience with IMAX ………………………… 38
　　Section 2：Reading ……………………………………………………………… 39
　　　　Multimedia ………………………………………………………………… 39
　　Section 3：Computer Terms …………………………………………………… 41
　　　　Do You Know? …………………………………………………………… 41
　　Section 4：Exercises …………………………………………………………… 42
　　Section 5：Furthering Reading ………………………………………………… 43
　　　　Adobe Photoshop ………………………………………………………… 43

Unit 6　Computer Networks …………………………………………………… 45

　　Section 1：Dialogue …………………………………………………………… 45
　　　　Computer Network Is Fascinating ……………………………………… 46
　　Section 2：Reading ……………………………………………………………… 47
　　　　What Is the Internet and How Does It Work? ……………………… 47
　　Section 3：Computer Terms …………………………………………………… 48
　　　　Do You Know? …………………………………………………………… 48
　　Section 4：Exercises …………………………………………………………… 50
　　Section 5：Furthering Reading ………………………………………………… 52
　　　　Mobile Internet …………………………………………………………… 52

Unit 7　Computer Security ……………………………………………………… 54

　　Section 1：Dialogue …………………………………………………………… 54
　　　　Safety Problems of a Computer ………………………………………… 54
　　Section 2：Reading ……………………………………………………………… 56

  Computer Security ……………………………………………………………… 56

 Section 3: Computer Terms ……………………………………………………… 59

  Do You Know? …………………………………………………………………… 59

 Section 4: Exercises ……………………………………………………………… 60

 Section 5: Furthering Reading …………………………………………………… 61

  Passwords Are Everywhere in Computer Security ……………………………… 61

Unit 8 E-commerce …………………………………………………………… 64

 Section 1: Dialogue ……………………………………………………………… 64

  Online Shopping and Some APPs ……………………………………………… 65

 Section 2: Reading ……………………………………………………………… 66

  E-commerce ……………………………………………………………………… 66

 Section 3: Computer Terms ……………………………………………………… 69

  Do You Know? …………………………………………………………………… 69

 Section 4: Exercises ……………………………………………………………… 70

 Section 5: Furthering Reading …………………………………………………… 72

  Dangdang Online Bookstore …………………………………………………… 72

Unit 9 New Technologies …………………………………………………… 74

 Section 1: Dialogue ……………………………………………………………… 74

  Cloud Computing ………………………………………………………………… 74

 Section 2: Reading ……………………………………………………………… 76

  Artificial Intelligence …………………………………………………………… 76

 Section 3: Computer Terms ……………………………………………………… 80

  Do You Know? …………………………………………………………………… 80

 Section 4: Exercises ……………………………………………………………… 81

 Section 5: Furthering Reading …………………………………………………… 82

  RealCine-virtual Reality for Everyone ………………………………………… 82

Unit 10 Job Application …………………………………………………… 85

 Section 1: Dialogues ……………………………………………………………… 86

  Working in an IT Company …………………………………………………… 86

 Section 2: Reading ……………………………………………………………… 87

 Section 3: Computer Terms ……………………………………………………… 89

  Do You Know? …………………………………………………………………… 89

Section 4: Exercises ·· 90
Section 5: Furthering Reading ··· 92
　　Resume ·· 93

参考译文 ··· 95

附录：计算机专业英语词汇表 ·· 103

附录：计算机专业英语专业术语词汇表 ································· 112

练习答案 ·· 116

Unit 1

What Is Computer English?

Unit Goals

In this unit, we are focusing on the following issues:
★ Introduction to Computer English
★ Why should we learn Computer English
★ What should we learn about Computer English
★ How should we learn Computer English

Introduction

  同学们，计算机技术在高速发展，网络应用在迅速普及，计算机专业方面的英语词汇被大量使用并不断出现。这种词汇的出现和更新速度之快、数量之多是任何一本辞典也难以做到及时收录的。每天都在产生新词，网络的另类新词在网络这个载体中川流不息。网络流行新词是你在网上经常看到却无法在辞典里找到的东西。所以，在计算机专业英语词汇发展如此迅猛的时代，了解计算机专业英语词汇的构成及语义特征，不但有助于我们知识的积累，开阔我们的视野，更有助于我们的英语学习和计算机学习。

  通常，在每次学习新知识之前，我们总要先搞清楚这样三个问题，也就是Why（为什么学）、What（学什么）以及How（如何学）。这三个问题我们将分别在本章的三个小节里加以说明。

  下面，就让我们来开始计算机专业英语的学习吧。

Section 1：Why

为什么要学习计算机专业英语（Why）

  同学们，当你选择计算机这个专业的时候，你就已经离不开英语的学习了。几乎所有的IT从业人员都知道学习计算机专业英语的重要性，但并不是所有人都具有学习的主动性。这个问题的提出就是要让我们从自身的需求出发，掌握学习的主动性。一旦你有强烈的学习动机，任何学习上的困难都不会让你屈服，而你的每一点进步都将给你带来无

比自豪的感觉。

为什么要学习计算机专业英语？用一句话来回答，就是：从事 IT 行业离不开英语。可以毫不夸张地说，英语是 IT 的行业语言。为什么这样说呢？

第一，计算机软件技术的更新越来越快，而这些技术大部分来源于英语国家。有关统计表明：计算机方面的论文 85% 以上是以英文形式发表的。等待译文会在很大程度上影响我们掌握新技术的时间。通常，一本外版计算机图书从获得版权到翻译出版要一年的时间。就算原作者消化新技术和写作的时间最短为一年，那么加起来这已经超过通常软件版本的更新周期了（一般为 1~2 年）。这意味着当你通过阅读翻译资料来掌握这一版本时，可能该版本已经被淘汰或至少将被淘汰了。

第二，在引进新技术时，往往受到语言障碍的制约，严重影响到对新技术的理解和消化。软件开发中的技术文档和资料大都用的是英文，即使有翻译好的，也经常晦涩难懂，或是译法混乱。例如：rollback 就有"回滚""回退""返回""重算"等多种译法。又如：看到"域"这一说法时，往往不清楚译者是根据"field""region"还是"domain"所译。而阅读原文则不存在这方面的问题。当然，这并不是说国内没有好的译者和译文，也不是说不能通过阅读译文来进行学习，而是说通过别人翻译来间接阅读的风险较大。如果能掌握计算机专业英语，利用第一手原文资料进行学习，除了效率之外，被误导的风险较小。

第三，计算机专业英语是一门专业基础课程，同时也是一门重要的工具课程。通过计算机专业英语相关内容的学习，学生可以在离开学校后，继续进行自我提高。这对于 IT 从业者来说，是非常必要的。我们在学校学习的知识通常在 3~5 年内就会被淘汰，而 IT 从业者的知识淘汰时间要更快更短。所以每一个 IT 从业者都面临离开学校后的自我学习、自我提高的问题。掌握一定的计算机专业英语知识，对于日后自学新知识和新技能会有较大的帮助。

同时，部分计算机专业考试中也涉及了计算机专业英语方面的内容。如程序员、软件设计师、网络管理员、网络工程师等考试中都有计算机专业英语的内容。掌握好本门课程对于同学们通过相关的考试也会有一定的帮助。同学们，由于英语的优势，印度、爱尔兰等国的软件业在国际上比我们更有竞争力。这并不是说我们的程序员在编程和开发能力上不如别人，而是在使用计算机专业英语水平上差距太大。曾经在南京举办过一次高规格的软件开发交流会，在会议上就遇到过印度专家讲课，英文翻译无法沟通的情景。因为太多的 IT 专用术语和缩略语以及很强的专业知识使得没有计算机背景的英语专业翻译无能为力，而在场的开发人员因为语言障碍又无法和印度专家直接沟通，因此错过了一次极好的交流学习机会。

现在，已经有越来越多的 IT 从业者意识到计算机专业英语的重要性了，这种压力一方面来自从业者本人进一步向高级程序员或资深 IT 开发人员发展的需要，另一方面来自后起之秀不断竞争的威胁。对于前者，没有较好的计算机专业英语，难以进一步发展，并晋升到更高的技术职位；对于后者，不少大学已经开始使用原版教科书进行专业授课，毕业生的计算机专业英语水平实在是后生可畏，挑战是不言而喻的。每一位立志在 IT 业有所发展的同学都应该掌握好计算机专业英语。

Section 2：What

计算机专业英语要学什么（What）

　　学什么呢，是单词，还是语法？其实都不是。计算机专业英语的学习是一项系统工程，需要找到一个适合自己的学习目标，并从词汇、语法、阅读、写作多方面去融会贯通。学习计算机专业英语是一个很宽泛的概念，若不根据个人的具体情况进行定义，恐怕难以有一个可以管理的学习目标。无目标的或边界不清的项目往往是失败的项目，在学计算机专业英语的问题上也是一样。

　　首先，学生根据自己的实际英语水平和以后的工作需要界定计算机专业英语学习的系统边界。我们学习计算机专业英语的根本目的是提高自己在英语环境中掌握计算机技术的能力，所以学习的内容应围绕本专业的领域展开。关于工作需要，我们可以大致将 IT 从业者划分成计算机研发人员、泛 IT 人员。其中计算机研发人员是指从事计算机研究和开发的专业人员，他们又划分为软件研发和硬件研发，显然程序员属于前者。计算机研发人员要掌握的计算机专业英语要求很专业而且要求高，但软件和硬件各有侧重。泛 IT 人员是指在 IT 行业从业的或与 IT 行业有密切联系的那些非研发人员，包括操作使用人员、技术管理人员、支持服务人员等。不难看出，泛 IT 人员对计算机专业英语的要求不是太高，也不太专业，一般能够使用英文界面的软件，能够阅读原版的操作手册和说明书即可。同学们应根据自己以后的职业规划，确定自己的学习目标。

　　其次，学生要先根据不同的工作需要界定不同的学习目标。针对以上的划分，我在下面大致给出了两种计算机专业英语的学习目标：①一般而言，对于立志在泛 IT 业中工作的同学来说，学习的目标应是掌握计算机专业英语的常用术语、缩略语；掌握计算机专业英语中语法和惯用法的表达方式和功能；能借助词典阅读英文文档和技术资料，阅读速度在 60 词/分钟以上；能使用英文编写简单的文档。②对于立志从事计算机研发的同学来说，学习的目标应订得高些：需掌握大量的计算机专业英语术语和缩略语；熟练掌握计算机专业英语中语法和惯用法的表达方式和功能；能阅读英文文档和技术资料，阅读速度在 100 词/分钟以上；能借助词典翻译专业技术图书；能使用英文编写简单的技术文档和程序注释等。

　　最后，在了解学习计算机专业英语的一般要求后，接下来就需要对自己的具体情况进行分析，制订一个学习计划或学习方向。同学们以前没有学过计算机专业英语，所以学习的重点应该是专业词汇和术语，了解计算机专业英语的一些规范译法和习惯用法。在有了一定的计算机专业英语知识之后，重点则应该是提高阅读速度和阅读质量，并逐渐习惯使用英文注释程序编写文档。在普通英语学习中，听、说、读、写、译是要求全面掌握的。因为不这样就无法掌握一门语言。而在计算机专业英语学习中，特别是在计算机专业英语的初级阶段的学习中，强调更多的是阅读和翻译的能力。也就是说，当你刚开始接触计算机专业英语的时候，只需要能读懂即可。比如你拿到了一本新的计算机软件的说明书，你能根据这个说明书很快地熟悉这个软件，或者你能在工具书的帮助下读懂这个说明书，从而会用这个软件，就可以了。所以，各位同学应该对自己有信心，不论你原来的英语水平如何，只要你能跟上我

们的学习进度，相信都会对计算机专业英语有较好的掌握。

 Section 3：How

如何学习计算机专业英语（How）

关于如何学习计算机专业英语，我们需注意以下几方面的问题。

1. 短期系统学习和长期日常学习相结合

我们在进行计算机专业英语学习的时候，应该注意把短期的系统学习和长期的日常学习结合起来。因为计算机专业英语的学习不同于考"托福"或考"GRE"，它无法通过短期突击来完成。计算机专业英语的不断发展现状也使我们必须坚持长期不断地学习。我们在学习和工作时应有意识地多接触和多使用计算机专业英语，边学边用、边用边学，不断积累、不断总结，才能将这门课程掌握好。

2. 选用一本适合于自己的教材或参考书

在系统学习的时候，我们还应根据自己的使用范围和学习要求选用一本适合自己的教材或参考书。目前计算机专业英语的教材多种多样，有偏硬件的，有偏软件的，有偏理论的（如数据结构、离散数学），有偏应用的（如软件工程、数据库开发），信息电子类的侧重于通信电子（如汇编语言、单片机）等。教材的选择主要从其难易程度、体系结构、易用性、专业侧重等方面选择。除以上需注意的方面外，还需注意教材内容的更新程度。计算机的发展日新月异，计算机专业英语也随之得到迅速发展。如果一本教材中的内容陈旧，那么就只能作为历史资料查阅，而不需要去专门学习了。以后大家走出校门，进行自我学习和提高的时候，也可以根据这里介绍的原则去选用一本适合于自己的参考书。

我们这本教材主要是从基础出发，让大家对计算机专业英语有一个全面的了解。大家可以在学习本教材后再根据自己的专业方向去选择提高。

3. 阅读原版资料

在日常生活中大量阅读英文书籍和资料是提高计算机专业英语水平的最佳途径。同学们在开始阅读原版资料时，会有"啃"的感觉，如果没有毅力的话，则难以读下去。一般在完整阅读2~3本原版书后，你就会发现读原版书的乐趣，进而可以享受"品"的感觉了。

4. 几个建议

为了让大家更好地学习计算机专业英语，在这里我给大家提几个可行性的做法。如果大家能够按照要求去做的话，那么大家就可以更快更好地学习计算机专业英语了。

（1）给自己起一个英文名字，并用作网名和登录名。
（2）保证本学期阅读一本原版的计算机资料，并坚持读完。
（3）上网尽量多访问英文的论述论坛和网站，不使用汉化的帮助。
（4）在程序中使用英文注释，不用中文或拼音作为变量名、字段名、文件名。

(5) 每天坚持记录和复习遇到的生词和术语。
(6) 对于缩略词，一定搞清每个字母的英文含义。
……

总之，要想学好计算机专业英语，需要注意下面四句话：词汇是基础、注意抓特点、不断多总结、原版多品味。

✉ Summary

同学们，在本章中，你们了解了有关计算机专业英语的入门知识，也就是三个问题：Why（为什么要学习计算机专业英语）、What（计算机专业英语要学什么）以及 How（如何学习计算机专业英语）。希望大家能在了解这三个基础问题后，树立学好计算机专业英语的信心，下定决心学好计算机专业英语，最后能完成计算机专业英语的学习任务。

Unit 2

Computer Hardware

> **Unit Goals**
>
> In this unit, we are focusing on the following issues:
> ★ Introduction to computer hardware
> ★ The components of computer hardware
> ★ The history of some hardware devices
> ★ Exercises about computer hardware

 Warm-up

Match the pictures with the words and phrases, then write the correct letter in the brackets to each.

( ) 1. modem ( ) 2. fax machine
( ) 3. notebook computer ( ) 4. desktop computer

A B C D

 Section 1: Dialogue

Directions: Read the following dialogue in pairs and talk about the computer hardware mentioned.
Situation: *Simon, a student from a vocational school is visiting a computer company. George, a computer programmer, is showing him around.*

Unit 2　Computer Hardware

What Is Computer Hardware?

Simon: Hello, George. I'm Simon, a vocational school student. This is my first time to visit a computer company. Would you please show me around?

George: Sure, Simon. Come over here. Let's start from the hardware.

Simon: Excuse me. What do you mean by hardware?

George: Well, it's used to describe computer devices. They are parts that you can touch. Some parts can be seen, but some parts cannot be seen from the outside. They are inside computer cases.

Simon: How many types of hardware are there?

George: That's a good question. There are generally four.

Simon: Then, what are they?

George: They are CPU, input devices, output devices and backing storage devices.

Simon: That's interesting. Could you tell me more?

George: Of course. Follow me and look at this computer on the desk here. Input devices, like the scanner, the floppy disk drive, the CD or DVD drive, the keyboard you used to type in, and the mouse you click. They pass information into the computer system.

Simon: And then, What is next?

George: The computer can store and process the information according to your instructions.

Simon: Wow, that's really amazing! Can I see how it works?

George: Now, Central Processing Unit, also known as CPU, processes everything from basic instructions to complex functions, and it's like the brain of the computer. CPU is attached on the motherboard inside, so you can't see unless we uninstall the computer.

Simon: Oh, I know, like the chips on circuit board.

George: That's right! And after processing, here come the output devices, which pass information out of the computer system, the content we see on the screen, the paperwork from the printer and the sound from the speaker, and so on.

Simon: What about the backing storage devices?

George: They store programs and data, such as hard disks and USB flash disks.

Simon: Thanks a lot!

📧 Words & Expressions

hardware [ˈhɑːdweə]　　　　　n. 计算机硬件
system [ˈsɪstəm]　　　　　　n. 体系，系统；制度
device [dɪˈvaɪs]　　　　　　n. 设备；装置
storage [ˈstɔːrɪdʒ]　　　　　n. 储藏；仓库；[计] 存储器
program [ˈprəʊɡræm]　　　　n. 节目；程序 v. 编制程序
data [ˈdeɪtə]　　　　　　　　n. (datum 的复数) 资料，材料；[计] 数据，资料
scanner [ˈskænə]　　　　　　n. 扫描仪

instruction[ɪnˈstrʌkʃ(ə)n]	n. 指令
process[prəˈses]	v. 处理
function[ˈfʌŋ(k)ʃ(ə)n]	n. 功能
motherboard[ˈmʌðəbɔːd]	n. 主板，主机板
uninstall[ʌnɪnˈstɔːl]	v. 卸载 n. 解除安装
floppy disk	n. 软盘
CPU（Central Processing Unit）	n. 中央处理器
computer case	计算机机箱
hard disk	硬盘
USB flash disk	U 盘

Role Play

A：Excuse me, Miss/Mrs/Mr... What is/are hardware/input devices/output devices/backing storage devices?

B：Well, it is/they are...

Section 2: Reading

Pre-reading Activities

1. Have you heard of the following computer parts? Talk about their main functions with your partner.

   CPU   memory   monitor
   RAM   ROM   printer

2. Do you have a computer? If yes, tell your peers about it.

The Components of Computer Hardware

When we talk about computer, such image (Figure 1) will appear in our mind: a display screen known as the basic output device, a keyboard usually together with a mouse known as the basic input device, and a machine box known as a cabinet.

Figure 1

Unit 2　Computer Hardware

Computer hardware can be divided into four categories: CPU, storage devices, input devices and output devices.

CPU

The central processing unit (CPU) is the brain of the computer. The design of the CPU affects the processing power and the speed of the computer as well as the amount of main memory it can use effectively.

Storage devices

We usually divide the storage devices into two types: the main memory and the secondary or auxiliary storage. The memory is divided into RAM (random access memory) and ROM (Read-only Memory). The main memory of most computers is composed of RAM. We can store data and programs into RAM. The amount of RAM you have in your PC directly affects the level of sophistication of the software you can use. When the computer is off, the main memory is empty. ROM can be read, but not be written. When the power is turned off, the instructions stored in ROM are not lost. Hard disk and USB Disk are common kinds of the secondary storage unit.

Input devices

The most common input devices are keyboard and mouse. The keys on a keyboard let you enter information and instructions into a computer. A mouse lets you select and move items on your screen.

Output devices

The most common output devices are monitor and printer. A monitor displays text and images. A printer produces a paper copy of information displayed on the screen. The ink jet printer and the laser printer are the two common printers at the current computer market.

✉ Words & Expressions

component [kəm'pəʊnənt]　　　　　n. 成分；零件 adj. 组成的
hardware ['hɑːdweə]　　　　　　　n. 硬件
software ['sɒf(t)weə]　　　　　　　n. 软件
desktop ['desktɒp]　　　　　　　　adj. 桌面的
modem ['məʊdem]　　　　　　　　n. 调制解调器
image ['ɪmɪdʒ]　　　　　　　　　　n. 图像
keyboard ['kiːbɔːd]　　　　　　　　n. 键盘
mouse [maʊs]　　　　　　　　　　n. 鼠标
cabinet ['kæbɪnət]　　　　　　　　n. 机箱；柜子
category ['kætəgəri]　　　　　　　n. 种类；类别
storage ['stɔːrɪdʒ]　　　　　　　　n. 存储器
compose [kəm'pəʊz]　　　　　　　v. 构成
random ['rændəm]　　　　　　　　adj. 随便的；任意的
access ['ækses]　　　　　　　　　n. 调取；存取
sophistication [səˌfɪstɪ'keɪʃn]　　　n. 复杂程度
monitor ['mɒnɪtə(r)]　　　　　　　n. 显示器

effective [ɪˈfektɪv]　　　　　　　　　　adj. 有效的；生效的

Basic Technical Terms

desktop computer	台式计算机
fax machine	传真机
display screen	显示屏
CPU（Central Processing Unit）	中央处理器
output device	输出设备
input device	输入设备
main memory	主存储器
secondary/auxiliary storage	辅助存储器
RAM（Random Access Memory）	随机存取存储器
ROM（Read-only Memory）	只读存储器
USB（Universal Serial Bus）Disk	U 盘

Complete the following sentences according to the passage.

1. The four categories of computer hardware are _____, _____, _____, _____.
2. A well designed CPU makes the computer have strong processing _____, high processing _____ and uses the amount of main _____ effectively.
3. A display screen known as the basic _____, and a keyboard usually together with a mouse as the basic _____.
4. The brain of computer is _____.
5. When the computer is off, _____ is empty.
6. Instruction and data are stored in _____.

Section 3：Computer Terms

Do You Know?

　　Computer hardware：计算机硬件，是指计算机系统中由电子、机械和光电元件等组成的各种物理装置的总称。从外观上来看，计算机由主机箱和外部设备组成。主机箱内主要包括CPU、内存、主板、硬盘驱动器、光盘驱动器、各种扩展卡、连接线、电源等；外部设备包括鼠标、键盘等。

　　CPU（Central Processing Unit）：中央处理器，是一块超大规模的集成电路，是一台计算机的运算核心（Core）和控制核心（Control Unit）。它的功能主要是解释计算机指令以及处理计算机软件中的数据。中央处理器主要包括运算器（算术逻辑运算单元，Arithmetic Logic Unit，ALU）和高速缓冲存储器（Cache）及实现它们之间联系的数据（Data）、控制及状态的总线（Bus）。它与内部存储器（Memory）和输入/输出（I/O）设备合称为电子计算机三大核心部件。

　　Memory：存储器，是现代信息技术中用于保存信息的记忆设备。计算机中全部信息，包括输入的原始数据、计算机程序、中间运行结果和最终运行结果都保存在存储器中。它根据控制器指定的位置存入和取出信息。存储器有若干个不同的类型：随机存取存储器（RAM）、只读存储器（ROM）、可编程只读存储器（PROM）、可擦可编程只读存储器（EPROM）、电可擦可编程只读存储器（EEPROM）等。

　　Bus：总线，是计算机各种功能部件之间传送信息的公共通信干线。按照计算机所传输的信息种类，计算机的总线可以划分为数据总线、地址总线和控制总线，分别用来传输数据、数据地址和控制信号。信息可以从多个源部件中的任何一个经总线传送到多个目标部件中的任意一个。

　　IBM（International Business Machines Corporation）：即国际商业机器公司，1911 年创立于美国。该公司创立时的主要业务为商业打字机，之后转为文字处理机，然后到计算机和有关服务。在过去的百年里，IBM 始终以超前的技术、出色的管理和独树一帜的产品领先于全球信息产业的发展。

　　ENIAC（Electronic Numerical Integrator And Computer）：埃尼阿克，即电子数字积分计算机，是世界上第一台现代通用计算机，诞生于 1946 年 2 月 14 日的美国宾夕法尼亚大学。虽然 ENIAC 体积庞大，耗电惊人，运算速度不过几千次，但它比当时已有的计算装置要快 1 000 倍，而且还有按事先编好的程序自动执行算术运算、逻辑运算和存储数据的功能。ENIAC 宣告了一个新时代的开始。

Section 4：Exercises

Ⅰ. **The following pictures show some hardware of computer. Please label the pictures with the given words and phrases.**

　　mouse _____　　　CPU _____　　　motherboard _____
　　printer _____　　　keyboard _____　　graphic card _____
　　memory _____　　hard disk _____　　monitor _____　　iPod _____

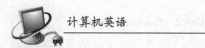

Ⅱ. **Fill in the blanks with some of the above mentioned words and phrases.**

Computer hardware consists of:

1. Input device: _____ is one of the common input devices.
2. Processor unit: It is divided into _____ and main memory.
3. Output device: _____ and _____ are the two most commonly used output devices.
4. Auxiliary storage unit: The common auxiliary storage devices are _____ and a hard disk drive.

Ⅲ. **Put the verbs in the passage into correct tenses.**

Some people _____ (say) that we live in the age of computers, but it _____ (be) also correctly described as the atomic age or the space age. Today, a journey from London to Cairo _____ (take) hours. Only a hundred years ago it _____ (take) weeks. Today, men _____ (think) seriously of going to Mars. Fifty years ago they only _____ (dream) about it. Today, we _____ (produce) energy by splitting the atom. A century ago, no one _____ (believe) it could be _____ (split). Technology _____ (advance) so quickly that cars and televisions _____ (be) out of date only a few years after they _____ (be) made.

Ⅳ. **Ability to explore.**

( ) 1. There is a notebook computer in an office, which works normally, but it cannot surf the web, the possible cause is _____.
　　A. network card failure
　　B. incorrect software setting
　　C. operating system failure
　　D. network card failure or incorrect software setting

( ) 2. In order to reduce your computer's power consumption on a hot summer day, what kinds of ways can we use to keep your computer cool? _____.
　　A. Don't use a CRT
　　B. Ensure that all the fans are functional
　　C. Manually adjust clock speeds, voltages via software
　　D. Manually adjust clock speeds, voltages via hardware
　　E. Turn off some useless programs
　　F. Toss out worn-out computers in your home

Ⅴ. **Fill in the table below by giving the corresponding translation with the help of dictionaries or Internet.**

English	Chinese
telecommunications industry	
	带宽

English	Chinese
information age	
	数码技术
electronics technology	
	营业执照
palm-sized computer	
	视听产品
download free and shared software	
	模拟信号

Section 5: Furthering Reading

Pre-reading Activity

Do you know anything about some "firsts" in computer history? Share your information with your classmates.

History of Computer Hardware

Since the early nineteenth century, the development of computers has seen numerous changes in the hardware. Here are some of the examples.

Mouse

In 1963 Douglas Engelbart invented the first mouse, which had two wheels set at a 90-degree angle to each other to keep track of the movement. The ball mouse was not invented until 1972. And the optical mouse was invented around 1980. It did not become popular until much later.

Laser printer

Gary Starkweather invented the first laser printer in 1969. However, it received wide use only after IBM introduced their branded laser printer known as IBM 3800 in 1976. It was as large as a room.

Floppy disks

Floppy disks were invented in 1970 and used till 1990s. However, today their use has become history.

Web server

The first web server was born in 1991. It was a NeXT workstation that Tim Berners-Lee used when he invented the World Wide Web at CERN. A note on the computer said, "This machine is a server. DO NOT POWER IT DOWN!" It meant that if you had shut it down in the early days you would have shut down the entire WWW.

Optical scanning

A device that can read text or illustrations printed on paper and translate the information into a form the computer can use. A scanner works by digitizing an image — dividing it into a grid of boxes and representing each box with either a zero or a one, depending on whether the box is filled in. (For color and gray scaling, the same principle applies, but each box is then represented by up to 24 bits.) The resulting matrix of bits, called a bit map, can then be stored in a file, displayed on a screen, and manipulated by programs.

Optical scanners do not distinguish text from illustrations; they represent all images as bit maps. Therefore, you cannot directly edit text that has been scanned. To edit text read by an optical scanner, you need an optical character recognition (OCR) system to translate the image into ASCII characters. Most optical scanners sold today come with OCR packages.

Answer the following questions according to the passage.

1. How did the first mouse look like?

2. When was the optical mouse invented?

3. When did the laser printer receive wide use?

4. Are floppy disks widely used today?

5. What special caution was given to the first web server?

Self-checklist

根据实际情况，从 A、B、C、D 中选择合适的答案：A 代表你能很好地完成该任务；B 代表你基本上可以完成该任务；C 代表你完成该任务有困难；D 代表你不能完成该任务。

A B C D

☐ ☐ ☐ ☐ 1. 理解并正确朗读听说部分的句子，正确掌握发音和语调。

☐ ☐ ☐ ☐ 2. 模仿对话部分的句型进行简单的对话。

☐ ☐ ☐ ☐ 3. 读懂本课的短文，并正确回答相关问题。

☐ ☐ ☐ ☐ 4. 向同学介绍计算机硬件及其功能等相关的基本知识。

☐ ☐ ☐ ☐ 5. 掌握并能运用本单元所学重点句型、词汇和短语。

☐ ☐ ☐ ☐ 6. 使用电子词典查阅 Furthering Reading，了解部分硬件的发展。

Unit 3

Operating System

Unit Goals

In this unit, we are focusing on the following issues:
★ The components of a computer system
★ The function of operating systems
★ The classifications of operating systems
★ The latest information about operating systems

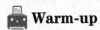 Warm-up

Look at the two architecture charts about Windows and Unix. Fill in the blanks with the Chinese equivalents.

```
A. 应用程序库    B. 图形设备接口    C. 内核接口    D. 实用程序
E. 编译程序      F. 视窗系统架构    G. 视窗系统可执行文件
H. 命令外壳程序  I. 应用程序        J. 视窗系统动态链接库
```

Windows Architecture

| Windows Executable File(.EXE) |
| Application Libraries(DLLs) |
| Windows DLLs(e.g. Kernel,User,GDI) |
| Windows Kernel & Low-level Drivers |

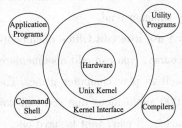
Unix Architecture

1. Windows Architecture _____   2. Windows Executable File _____
3. Application Libraries _____   4. Windows DLLs _____
5. GDI _____   6. Kernel Interface _____
7. Application Programs _____   8. Utility Programs _____
9. Compilers _____   10. Command Shell _____

Section 1: Dialogue

Directions: Read the following dialogue in pairs and talk about the operating system mentioned.

Situation: *All computers must use operating systems. Different people have different likes and dislikes about the operating systems they use. Simon, a student majoring in computer, wants to buy a personal computer, so he asks his friend George, a computer expert, to give him some advice on operating systems.*

What Operating System Do You Use?

Simon: Hi, George. I have to work on the computer every day both in class and after class, and it seems it's become a must-have tech product nowadays.

George: Yes, Simon. Do you plan to buy one?

Simon: Well, I really need a personal computer, but I felt it may be too expensive for me. Can you recommend some PCs that I can afford?

George: Eh, there are many kinds of PCs, such as desktops, laptops and tablets. I assembled my first desktop 10 years ago and last summer I replaced it with another laptop at about 4,000 yuan. Last month, I bought a tablet for my daughter, which is much cheaper.

Simon: Wow, I think a tablet might be suitable for me.

George: That's right! And the Android Tablet is a great deal more cost-effective than the iPad or a laptop. There are lots of Android Tablets that you can buy from the market at much lower prices. For example, you can simply get a Huawei Tablet PC running Android OS in 800 to 1,500 yuan range that will offer you identical performance to the iPad.

Simon: That sounds very attractive to me.

George: As a matter of fact, some Android Tablet PCs are actually working more rapidly than the very common iPad.

Simon: Can I use it to edit Office documents?

George: Of course, you can edit documents in MS Word, Excel, PowerPoint and read PDF files, as well as play computer games. The Android OS is an open source; it has more than 70,000 applications that you can download directly to your device.

Simon: Amazing, I can't wait to have one.

George: Go ahead, and the Android OS was already installed when you bought it. So the moment you open it, you can experience the speed and fluency of the Android OS.

✉ Words & Expressions

must-have	*n.*	必需品
desktop ['desktɒp]	*n.*	台式机

laptop [ˈlæptɒp]	n.	笔记本计算机
tablet [ˈtæblɪt]	n.	平板计算机
assemble [əˈsemb(ə)l]	v.	组装
Android [ˈændrɔɪd]	n.	基于 Linux 平台的开源手机操作系统
cost-effective [ˈkɔːstəˈfektɪv]	adj.	有成本效益的，划算的；合算的
OS (operating system)	n.	操作系统
identical [aɪˈdentɪk(ə)l]	adj.	完全相同的；同一的
performance [pəˈfɔːm(ə)ns]	n.	性能
device [dɪˈvaɪs]	n.	设备；装置；终端
fluency [ˈfluːənsɪ]	n.	流畅；流畅性

Role Play

A: Excuse me, Miss/Mrs/Mr... What is your operating system?
B: Well, it is...

Section 2: Reading

Pre-reading Activities

Some of the following pictures stand for operating systems. Pick them out.

1 (　)　　2 (　)　　3 (　)　　4 (　)

5 (　)　　6 (　)　　7 (　)

Operating System

As we all know, a computer system can be roughly divided into 4 components (Figure 1).

The hardware provides the basic computing resources. The application programs define the ways in which these resources are used to solve the computing problems of the users. There may be many different users trying to solve different problems, so there may be many different application programs.

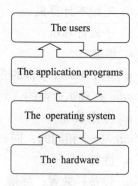

Figure 1

An operating system (OS) is system software that manages computer hardware and software resources and provides common services for computer programs. It acts as an interface between a user of a computer and the computer hardware. It provides an environment in which a user may execute programs. The primary goal of an operating system is thus to make the computer system convenient to use. A secondary goal is to use the computer hardware in an efficient way. We can view an operating system as a resource allocator. A computer system has many resources which may be required to solve a problem: CPU time, memory space, file storage, input/output (I/O) devices, and so on.

There are a wide variety of operating systems that can be installed on various devices. According to the categories of applications, OS falls into three major classifications: desktop operating system, server operating system and embedded operating system.

Desktop operating system

Desktop operating systems are primarily used on personal computers. In terms of hardware architecture, the PC market is mainly divided into two camps: PC and Mac. From the perspective of software, it can be mainly divided into two categories: Unix-like operating system and Windows operating system:

1. Unix and Unix-like OS: Mac OS X, Linux distribution (Debian, Ubuntu, Linux Mint, openSUSE, Fedora, etc.)

2. Microsoft Windows OS: Windows XP, Windows Vista, Windows 7, Windows 8, etc.

Server operating system

A server operating system generally refers to an operating system installed on a mainframe computer, such as a Web server, an application server, a database server, and so on. Server operating systems are mainly concentrated in three categories:

1. Unix series: SUNSolaris, IBM-AIX, HP-UX, FreeBSD, etc.

2. Linux series: Red Hat Linux, CentOS, Debian, Ubuntu, etc.

3. Windows series: Windows Server 2003, Windows Server 2008, Windows Server 2008 R2, etc.

Embedded operating system

An embedded operating system is the operating system applied in embedded systems.

Embedded systems are widely used in all aspects of life, ranging from portable devices to large fixed facilities, such as digital cameras, mobile phones, tablet computers, household appliances, medical equipment, traffic lights, avionics and factory control equipment.

Operating systems commonly used in the embedded field include Embedded Linux, Windows Embedded, VxWorks, etc., as well as operating systems widely used in smart phones or tablets and other electronic products, such as Android, iOS, Symbian, Windows Phone and BlackBerry OS, etc.

Words & Expressions

roughly['rʌfli]	*adv.*	大约,大致
component[kəm'pəunənt]	*n.*	部分；成分
define[dɪ'faɪn]	*vt.*	定义
interface['ɪntəfeɪs]	*n.*	接口,交界面
coordinate[kəu'ɔːdɪneɪt]	*vt.*	协调
purpose['pɜːpəs]	*n.*	目的
execute['eksɪkjuːt]	*v.*	执行；完成；履行
primary['praɪm(ə)rɪ]	*adj.*	主要的；初级的；基本的
convenient[kən'viːnɪənt]	*adj.*	方便的
secondary['sek(ə)nd(ə)rɪ]	*adj.*	次要的；第二的
efficient[ɪ'fɪʃnt]	*adj.*	有效的；高效率的
allocator['æləukeɪtə]	*n.*	分配器
variety[və'raɪətɪ]	*n.*	多样；种类
device[dɪ'vaɪs]	*n.*	设备；装置；终端
category['kætɪg(ə)rɪ]	*n.*	种类,分类；范畴
classification[ˌklæsɪfɪ'keɪʃn]	*n.*	分类,类别；等级
embedded[ɪm'bedɪd]	*adj.*	嵌入式的,植入的,内含的
architecture['ɑːkɪtektʃə]	*n.*	建筑；架构
camp[kæmp]	*n.*	露营；营地；阵营
perspective[pə'spektɪv]	*n.*	观点；远景
mainframe['meɪnfreɪm]	*n.*	主机；大型机
database['deɪtəbeɪs]	*n.*	数据库,资料库
concentrate['kɒns(ə)ntreɪt]	*vt.*	集中,专注于
series['sɪəriːz]	*n.*	系列；连续
aspect['æspekt]	*n.*	方面；方向
portable['pɔːtəb(ə)l]	*adj.*	手提的；便携式的
facility[fə'sɪlətɪ]	*n.*	工具；设备
digital['dɪdʒɪt(ə)l]	*adj.*	数字的；数码的
tablet['tæblɪt]	*n.*	平板计算机
equipment[ɪ'kwɪpm(ə)nt]	*n.*	设备；器材
avionics[ˌeɪvɪ'ɒnɪks]	*n.*	航空电子设备
environment[ɪn'vaɪrənmənt]	*n.*	环境

Basic Technical Terms

operating system	操作系统
application program	应用程序
hardware resources	硬件资源
resource allocator	资源分配器
CPU time	时序
memory space	储存空间
file storage space	文件储存空间
input/output (I/O) device	输入输出设备
Desktop OS	桌面操作系统
hardware architecture	硬件架构
Unix-like OS	类 Unix 操作系统
Linux distribution	Linux 发行版
Server OS	服务器操作系统
mainframe computer	大型计算机
Embedded OS	嵌入式操作系统
digital camera	数码相机
tablet computer	平板计算机
household appliance	家用电器
embedded Linux	嵌入式 Linux

Tell whether the following statements are true (T) or false (F).

(   ) 1. A computer system can be roughly divided into 4 parts: the hardware, the software, the application programs and the users.

(   ) 2. An operating system can be viewed as a control program.

(   ) 3. The operating system provides the basic computing resources.

(   ) 4. With an operating system, it is convenient for the programmers to perform some very complex tasks of programming.

(   ) 5. Windows XP, Windows Vista, Windows 7, Windows 8 of Microsoft are in the group of Server OS.

Section 3: Computer Terms

Do You Know?

Unix Operating System：Unix 操作系统，是一个强大的多用户、多任务操作系统，支持多种处理器架构；按照操作系统的分类，它属于分时操作系统，最早由 Ken Thompson、

Dennis Ritchie 和 Douglas McIlroy 于 1969 年在 AT&T 的贝尔实验室开发。该系统采用树状目录结构，具有良好的安全性、保密性和可维护性。Unix 系统大部分是由 C 语言编写的，这使得系统易读、易修改、易移植。它提供了丰富的、精心挑选的系统调用，整个系统的实现十分紧凑、简洁。

Linux Operating System：Linux 操作系统，是一套免费使用和自由传播的类 Unix 操作系统，是一个基于 Posix 和 Unix 的多用户、多任务、支持多线程和多 CPU 的操作系统。它能运行主要的 Unix 工具软件、应用程序和网络协议。它支持 32 位和 64 位硬件。Linux 继承了 Unix 以网络为核心的设计思想，是一个性能稳定的多用户网络操作系统。

Novell Operating System：Novell 操作系统，是 Novell 公司的产品，被认为是网络服务器操作系统的鼻祖，其产品早在 20 世纪 80 年代进入中国。Novell Inc. 是世界上最具实力的网络系统公司，曾经在全世界软件行业中排名第五。

OS/2 Operating System：OS/2 操作系统，是由微软和 IBM 公司共同创造的、后来由 IBM 单独开发的一套操作系统。OS/2 是"Operating System/2"的缩写，是因为该系统作为 IBM 第二代个人电脑 PS/2 系统产品线的理想操作系统引入的。在 DOS 于 PC 上取得巨大成功后，在 GUI 图形化界面的潮流影响下，IBM 和 Microsoft 共同研制和推出了 OS/2 这一当时先进的个人计算机上的新一代操作系统。最初它主要是由 Microsoft 开发，由于在很多方面的差别，微软最终放弃了 OS/2 而转向开发 Windows "视窗"系统。

Mac OS Operating System：Mac OS 操作系统，是苹果公司为 Mac 系列产品开发的专属操作系统。Mac OS 是苹果 Mac 系列产品的预装系统，处处体现着简洁的宗旨。Mac OS 是全世界第一个基于 FreeBSD 系统采用"面向对象操作系统"的、全面的操作系统。"面向对象操作系统"是史蒂夫·乔布斯（Steve Jobs）于 1985 年被迫离开苹果后成立的 NeXT 公司所开发的。后来苹果公司收购了 NeXT 公司。史蒂夫·乔布斯重新担任苹果公司 CEO，Mac 开始使用的 Mac OS 系统得以整合到 NeXT 公司开发的 Openstep 系统上。现在最新的正式版本是 Mac OS Mojave。

Windows XP Operating System：Windows XP 操作系统，是美国微软公司研发的基于 X86、X64 架构的 PC 和平板计算机使用的操作系统，于 2001 年 8 月 24 日发布 RTM 版本，并于 2001 年 10 月 25 日开始零售。其名字中"XP"的意思来自英文中的"体验（Experience）"。该系统是继 Windows 2000 及 Windows ME 之后的下一代 Windows 操作系统，也是微软首个面向消费者且使用 Windows NT5.1 架构的操作系统。

Windows 7 Operating System：Windows 7 操作系统，是由微软公司（Microsoft）开发的操作系统，内核版本号为 Windows NT 6.1。Windows 7 可供家庭及商业工作环境：笔记本计算机、平板计算机、多媒体中心等使用。和同为 NT6 成员的 Windows Vista 一脉相承，Windows 7 继承了包括 Aero 风格等多项功能，并且在此基础上增添了一些功能。Windows 7 可供选择的版本有：初级版（Starter）、家庭普通版（Home Basic）、家庭高级版（Home Premium）、专业版（Professional）、企业版（Enterprise）（非零售）、旗舰版（Ultimate）。2009 年 7 月 14 日，Windows 7 正式开发完成，并于同年 10 月 22 日正式发布，10 月 23 日，微软于中国正式发布 Windows 7。2015 年 1 月 13 日，微软正式终止了对 Windows 7 的主流支持，但仍然继续为 Windows 7 提供安全补丁支持，直到 2020 年 1 月 14 日正式结束对 Windows 7 的所有技术支持。

Section 4: Exercises

Ⅰ. Fill the blanks with the words or phrases of the text.

1. A computer system can be roughly divided into 4 _____.
2. The hardware provides basic computing _____.
3. An Operating System (OS) acts as an _____ between a user of a computer and the computer hardware.
4. We can view an operating system as a resource _____.
5. OS falls into three major classifications: _____, _____ and _____.

Ⅱ. Fill the blanks with the following words or phrases. Change the form if necessary.

act as	refer to	manage	a wide variety of
divide into	provide	primary	in an efficient way

1. When we say "diggers," we _____ any outside underground animal.
2. I wonder if they'd _____ a hyperlink (超链接) to our site.
3. Ninety-nine percent of _____ pupils now have hands-on experience of computers.
4. Its large size makes the park more difficult to _____.
5. You can choose what you like from _____ electronic products.
6. Could you do me a favor to tell me how to finish the assignment _____?
7. Mr. Li _____ group leader while Mr. Zhang was ill.
8. The large birthday cake _____ several pieces.

Ⅲ. Translate the following sentences into Chinese.

1. A computer system can be roughly divided into 4 components, the hardware, the operating system, the application programs, the users.

2. There may be many different users trying to solve different problems, so there may be many different application programs.

3. An operating system (OS) is system software that manages computer hardware and software resources and provides common services for computer programs.

4. A computer system has many resources which may be required to solve a problem: CPU time, memory space, file storage space, input/output (I/O) devices, and so on.

5. Embedded systems are widely used in all aspects of life, ranging from portable devices to large fixed facilities, such as digital cameras, mobile phones, tablet computers, household appliances, medical equipment, traffic lights, avionics and factory control equipment.

Ⅳ. **Ability to explore**.

Why are operating systems considered inseparable from the hardware? Which of the following reasons is not true? _____
A. Operating systems can manage the basic hardware resources.
B. Operating systems can solve all the special problems of users.
C. Operating systems can provide a more hospitable interface to users.
D. Operating systems can make computer operation efficient.

Ⅴ. **Fill in the table below by giving the corresponding translation with the help of dictionaries or Internet**.

English	Chinese
software package	
	文档管理
fragmentation	
	影像采集
text formatting	
	交互式处理
hardware compression	
	服务器系统

 Section 5: Furthering Reading

Android

Android is a Linux-based operating system developed by the Open Handset Alliance, a group that includes Google and more than 30 technology and mobile companies. The most widely used mobile operating system in the world, Android was built from the ground up with current mobile device capability in mind, which enables developers to create mobile applications that take full advantage of all the features a mobile device has to offer. It is an open platform, so anyone can download and use Android, although hardware manufacturers must adhere to certain specifications in order to be called "Android compatible." A variety of manufacturers produce devices that run the Android operating system, adding their own interface elements and bundled software. As a result, an Android smartphone manufactured by Samsung may have different user interface features from one manufactured by Google.

Features unique to recent versions of the Android operating system include the following:
- Google Play APP Store provides access to APPs, songs, books, and movies.
- Google Drive provides access to e-mail, contacts, calendar, photo files, and more.
- Face recognition or fingerprint scanner can unlock the device.
- Share contacts and other information by touching two devices together (using NFC technology).
- Speech output assists users with vision impairments.
- Voice recognition capability enables users to speak instructions.
- Built-in heart rate monitor works with phone APPs.

Choose the right answers according to the passage.

1. Which kind of platform is Android according to this article?
   A. Close.　　　　B. Open.　　　　C. Obsolete.　　　　D. Old.
2. Do hardware manufacturers have to adhere to certain specifications in order to be called "Android compatible" according to this article?
   A. Yes.　　　　B. No.　　　　C. Not necessary.　　　　D. None of the above.
3. Which kind of operating system is a Samsung smartphone using according to this article?
   A. iOS.　　　　B. Windows Phone.　　　　C. Android.　　　　D. Blackberry.

Self-checklist

根据实际情况，从 A、B、C、D 中选择合适的答案：A 代表你能很好地完成该任务；B 代表你基本上可以完成该任务；C 代表你完成该任务有困难；D 代表你不能完成该任务。

A	B	C	D	
□	□	□	□	1. 理解并正确朗读听说部分的句子，正确掌握发音和语调。
□	□	□	□	2. 模仿听说部分的句型进行简单的对话。
□	□	□	□	3. 读懂本课的短文，并正确回答相关问题。
□	□	□	□	4. 向同学介绍计算机各操作系统及其特征等相关的基本知识。
□	□	□	□	5. 掌握并能运用本单元所学重点句型、词汇和短语。
□	□	□	□	6. 使用电子词典查阅 Furthering Reading，了解软件的最新信息。

Unit 4

Application Software

Unit Goals

In this unit, we are focusing on the following issues:
★ Identify the categories of application software
★ Identify the key features of widely used software
★ Describe the function of several utility programs
★ Know the latest information about application software

Warm-up

Ⅰ. Tick out the general office automatic devices.

(   ) facsimile machine   (   ) duplicator
(·) printer   (   ) microcomputer
(   ) scanner   (   ) digital camera
(   ) projecting apparatus   (   ) recorder
(   ) group telephone   (   ) UPS power source
(   ) all-in-one multifunctional machine   (   ) shredder

Ⅱ. The general office automatic devices enable you to save _____.
A. on space   B. on cabling
C. on power requirements   D. on consumables

Section 1: Dialogue

Directions: Read the following dialogue in pairs and talk about the document compatibility of different versions of Microsoft Office.

Situation: *Different people use a word processor to do different things. Simon, a clerk in a foreign trade company, cannot open the ". docx" file with his office program and asks George, a computer expert, to give him a hand.*

— 25 —

How to Solve an Office Problem?

Simon: Hi, George. Are you available right at the moment?

George: Yep, I just finished my job.

Simon: Can you do me a favor?

George: Of course. What's up?

Simon: Recently, I've often received some new documents from clients with filename extension ". docx." I can't open or edit the documents with my Microsoft Office. How can I solve this problem?

George: Which version of Microsoft Office do you use?

Simon: Microsoft Office 2003.

George: No wonder you had that problem, for ". docx" is the new filename extension that Microsoft Word 2007 uses when it saves the documents as the default file format. So you can install Microsoft Office 2007 to open and edit those documents.

Simon: Oh, but that's too complicated. Is there any other way?

George: If you download and install an Office compatibility pack from Microsoft, it will enable you to open and edit ". docx" files. However, some objects generated from new features in Microsoft Word 2007 will be converted to images, so you can only view but not edit them.

Simon: That's very helpful. Thank you so much!

George: You're welcome. Don't hesitate to ask me if you have any other questions.

Simon: I won't. Thank you very much.

✉ Words & Expressions

available [əˈveɪləb(ə)l]   *adj.* 有空的
document [ˈdɒkjʊm(ə)nt]   *n.* 文档
filename extension [ˈfaɪlnem] [ɪkˈstenʃ(ə)n]   *n.* 文件扩展名
default [dɪˈfɔːlt]   *n.* 默认
format [ˈfɔːmæt]   *n.* 格式
install [ɪnˈstɔːl]   *v.* 安装
compatibility pack [kəmˌpætɪˈbɪlɪtɪ] [pæk]   *n.* 兼容包
convert [kənˈvɜːt]   *v.* 转变

🐓 Role Play

A: Excuse me, Miss/Mrs/Mr... Which version of Microsoft Office do you use?

B: Well, I use...

Unit 4 Application Software

 Section 2: Reading

Pre-reading Activity

The pictures below show different toolbars. Choose the correct explanation for each toolbar. Write the letter A-H in the brackets.

1.( ) 2.( ) 3.( ) 4.( ) 5.( ) 6.( ) 7.( ) 8.( )

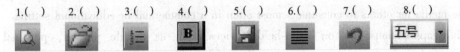

A. Change the point size of the selected text
B. Bold the characters in the selected text
C. Undo the last document change
D. Justify the selected paragraph
E. Show what the printed pages of document will look like
F. Open an existing document
G. Save the active document on disk
H. Number the selected text

Application Software

Software is another name for program. As we all know, a computer can do nothing without the support of software. If we describe a computer as a person, we usually say that hardware is like a person's body, while the software is like the soul. Just like a person has a lot of ideas, there are several kinds of software, each of which does different jobs. Software can be divided into two major categories: system software and application software. Without the former, your computer won't run. And without the latter, your computer — no matter how powerful — won't do much to help you run your business.

With system software, you get your computer going when you turn it on: writing information to a disk, checking for viruses and a host of other activities. Application software may be customer-made or packaged. It does end-users work.

There are many types of application software:

An application suite consists of multiple applications bundled together. They usually have related functions, features and user interfaces, and may be able to interact with each other, e. g. open each other's files.

Business applications often come in suites, e. g. Microsoft Office, OpenOffice. org, and iWork, which bundle together a word processor, a spreadsheet, etc. ; but suites exist for other purposes, e. g. graphics or music.

Enterprise software addresses the needs of organization processes and data flow, often in a large distributed environment.

Information worker software addresses the needs of individuals to create and manage information.

Educational software is related to content access software, but has the content and/or features adapted for use by educators or students.

Simulation software are computer software for simulation of physical or abstract systems for either research, training or entertainment purposes.

Media development software addresses the needs of individuals who generate print and electronic media for others to consume, most often in a commercial or educational setting.

Mobile applications run on hand-held devices such as mobile phones, personal digital assistants, and enterprise digital assistants.

Product engineering software is used in developing hardware and software products. This includes computer aided design (CAD), computer aided engineering (CAE), computer language editing and compiling tools, integrated development environments, and application programmer interfaces.

Words & Expressions

process [ˈprəuses]	v.	处理
database [ˈdeɪtəbeɪs]	n.	数据
compiler [kəmˈpaɪlə]	n.	编译程序
spreadsheet [ˈspredʃiːt]	n.	表格
application [ˌæplɪˈkeɪʃ(ə)n]	n.	应用
former [ˈfɔːmə]	adj.	前者的
latter [ˈlætə]	adj.	后者的
virus [ˈvaɪrəs]	n.	病毒
package [ˈpækɪdʒ]	n.	包
integrated [ˌɪntɪˈgreɪtɪd]	adj.	集成的
generate [ˈdʒenəreɪt]	v.	产生
attempt [əˈtem(p)t]	v.	企图
publish [ˈpʌblɪʃ]	v.	出版；印刷
architect [ˈɑːkɪtekt]	n.	建筑师
graphics [ˈgræfɪks]	n.	图表
create [kriːˈeɪt]	v.	引起；造成；创作
edit [ˈedɪt]	v.	编辑
click [klɪk]	v.	点击；敲击
assign [əˈsaɪn]	v.	分配；指定
folder [ˈfəuldə]	n.	文件夹
insert [ɪnˈsɜːt]	v.	插入

Unit 4 Application Software

exist [ɪgˈzɪst]	v. 存在
multiple [ˈmʌltɪpl]	adj. 多重的；复杂的；多功能的
interface [ˈɪntəfeɪs]	n. 界面；交界面
enterprise [ˈentəpraɪz]	n. 企（事）业单位；事业
simulation [ˌsɪmjuˈleɪʃn]	n. 模仿；模拟
consume [kənˈsjuːm]	v. 消耗；消费；耗尽；毁灭
digital [ˈdɪdʒɪt(ə)l]	adj. 数字的；数据的
compile [kəmˈpaɪl]	v. 编译；编制
integrate [ˈɪntɪgreɪt]	v. 使一体化；使整合

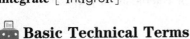 **Basic Technical Terms**

operating system	操作系统
system software	系统软件
application software	应用软件
database management systems (DBMS)	数据库管理系统
integrated software	集成软件
commercial software package	商业软件包
desktop publishing software	桌面印刷软件
graphics software	图形软件
file name text	文件名文本框
print layout view	打印预览
target diskette	目标磁盘
user interface	用户界面
application suite	成套应用软件
related function	相关函数
interact with	与……相互作用
Microsoft Office	微软办公软件
word processor	文字处理器
enterprise software	企业软件
educational software	教育软件
distributed environment	分布环境
simulation software	仿真软件
media development	媒体开发软件
mobile application software	移动应用软件
product engineering software	产品工程软件
computer aided design (CAD)	计算机辅助设计
computer aided engineering (CAE)	计算机辅助工程

Mark the following sentences with true or false according to the passage.
(   ) 1. The computer hardware is the basement of the software.
(   ) 2. A computer can do anything without the support of software.
(   ) 3. Software can be divided into application software and operating system software.
(   ) 4. We can get application software from the Internet.
(   ) 5. DBMS is a system software.
(   ) 6. An application suite must interact with each other.
(   ) 7. Business applications often come in suites.
(   ) 8. Educational software is for either research, training or entertainment purposes.

Section 3: Computer Terms

Do You Know?

　　Computer Hardware：计算机硬件，是计算机系统中由电子、机械和光电元件等组成的。Application Software 应用软件是和系统软件相对应的，是用户可以使用的各种程序设计语言，以及用各种程序设计语言编制的应用程序的集合，分为应用软件包和用户程序。应用软件可以拓宽计算机系统的应用领域，放大硬件的功能。

　　Database Management System：数据库管理系统，是一种操纵和管理数据库的大型软件，用于建立、使用和维护数据库，简称 DBMS。它对数据库进行统一的管理和控制，以保证数据库的安全性和完整性。用户通过 DBMS 访问数据库中的数据，数据库管理员也通过 DBMS 进行数据库的维护工作。它可使多个应用程序和用户用不同的方法在同时或不同时刻去建立、修改和询问数据库。大部分 DBMS 提供数据定义语言 DDL（Data Definition Language）和数据操作语言 DML（Data Manipulation Language），供用户定义数据库的模式结构与权限约束，实现对数据的追加、删除等操作。

　　Microsoft Office：微软公司开发的办公软件，为 Microsoft Windows 和 Mac OS X 而开发。与办公室应用程序一样，它包括联合的服务器和基于互联网的服务。该软件最初出现于 20 世纪 90 年代早期，最初是一个推广名称，指一些以前曾单独发售的软件的合集。当时主要的推广重点是购买合集比单独购买要省很多钱。最初的 Office 版本只有 Word、Excel 和 PowerPoint；另外一个专业版包含 Microsoft Access；随着时间的流逝，Office 应用程序逐渐整合，共享一些特性，例如拼写和语法检查、OLE 数据整合和微软 Microsoft VBA（Visual Basic for Applications）脚本语言。

　　Microsoft Office Word：文字处理软件，被认为是 Office 的主要程序。它在文字处理软件市场上拥有统治份额。它私有的 DOC 格式被尊为一个行业的标准，尽管它的新版本 Word 2007 也支持一个基于 XML 的格式。它适宜于 Windows 和 Mac 平台。它的主要竞争者是 Writer、Star Office、Corel WordPerfect 和 Apple Pages。

　　Microsoft Office PowerPoint：微软公司设计的演示文稿软件。用户不仅可以在投影仪或者计算机上进行演示，也可以将演示文稿打印出来，制作成胶片，以便应用到更广泛的领域中。利用PowerPoint不仅可以创建演示文稿，还可以在互联网上召开面对面会议、远程会议或在网上给观众展示演示文稿。PowerPoint做出来的东西叫演示文稿，它是一个文件，其格式后缀名为".ppt"，或者也可以保存为".pdf""图片格式"等，2010年和2013年版本中可保存为视频格式。演示文稿中的每一页叫作幻灯片，每张幻灯片都是演示文稿中既相互独立又相互联系的内容。

Microsoft Office Excel：电子数据表程序（进行数字和预算运算的软件程序），是最早的Office组件。Excel内置了多种函数，可以对大量数据进行分类、排序甚至绘制图表等。像Microsoft Office Word，它在市场上拥有统治份额。它适宜于Windows和Mac平台。它的主要竞争者是Calc、Star Office和Corel Quattro Pro。

Microsoft Office Outlook：个人信息管理程序和电子邮件通信软件，在Office 97版接任Microsoft Mail。但它与系统自带的Outlook Express是不同的：它包括一个电子邮件客户端、日历、任务管理者和地址本——它比Outlook Express的功能多得多。

Microsoft Office FrontPage：微软公司推出的一款网页设计、制作、发布、管理的软件，由于具有良好的易用性而被认为是优秀的网页初学者的工具。但其功能无法满足更高要求，所以在高端用户中，大多数使用Adobe Dreamweaver作为代替品。它的主要竞争者也是Adobe Dreamweaver。

Section 4：Exercises

Ⅰ. **Why are operating systems considered inseparable from the hardware? Which of the following reasons is not true?**

    A. Operating systems can manage the basic hardware resources.

    B. Operating systems can solve all the special problems of users.

    C. Operating systems can provide a more hospitable interface to users.

    D. Operating systems can make computer operation efficient.

Ⅱ. **Complete the sentences below with the correct verbs.**

1. Print Layout View _____ (shows/clicks) how your page will look on paper.
2. In the File Name Text, _____ (show/enter) the name you want to assign to the document file.
3. _____ (Save/Double-click) the document in a different folder or drive.
4. Please _____ (insert/save) a new paragraph in your essay, then rename your file.
5. File already _____ (exists/clicks).
6. A lot of software in your computer is out of date, you must _____ (download/upgrade) it.
7. Your first program is ready to _____ (compile/ translate) and run.
8. Teachers have a limited amount of time to _____ (contact/interact) with each child.
9. There were officials to whom he could _____ (relate /connect) the whole story.
10. Some good suggestions will be able to _____ (integrate/link) the plan.

Ⅲ. **Fill in the blanks according to the passage.**

1. Another name for program is _____ in computer science.
2. Software can be divided into _____ and _____ .
3. The functions of _____ software enable you to write information, check for viruses and many other activities when you turn your computer on.
4. People can use _____ programs to handle specific types of information and achieve useful result, such as cost analysis, real estate management, and so on.
5. An application suite consists of _____ bundled together. They usually have related _____ , _____ and _____ .
6. Educational software has the content and/or features adapted for use by _____ or _____ .
7. Mobile phones, one kind of _____ applications, run on hand-held devices such as mobile phones.
8. Product engineering software is used in developing _____ and _____ products.

Ⅳ. **Reading comprehension.**

Microsoft Office

The commonly used software in office automation include Microsoft, Corel and KingSoft. Microsoft Office is a commonly-used office suite of inter-related desktop applications, servers and services for the Microsoft Windows and Mac OS X operating systems, introduced by Microsoft in 1989. The first version of Office contained Microsoft Word, Microsoft Excel and Microsoft PowerPoint. Over the years, Office applications have grown closer with shared features. The current versions are Office 2010 for Windows, released on June 15, 2010, and Office 2011 for Mac OS X, released on October 26, 2010.

Word

Microsoft Word is very powerful. It works very well on the platform of Windows. It realizes MYSIWYG (What you see is what you get) feature. Microsoft Word is a word processor and was previously considered to be the main program in Office. The first version of Word, released in the autumn of 1983, was for the MS-DOS operating system and had the distinction (区别；分清) of introducing the mouse to a broad population.

Excel

Microsoft Excel is a spreadsheet program which originally competed with the dominant (支配的，主要的) Lotus 1-2-3, but eventually outsold it. Microsoft released the first version of Excel for the Mac in 1985, and the first Windows version in November, 1987.

PowerPoint

Microsoft PowerPoint is a popular presentation program for Windows and Mac. It is used to create slideshows, composed of text, graphics, movies and other objects, which can be displayed on-screen and navigated through by the presenter or printed out on transparencies (透明；透明度)

or slides.

Outlook/Entourage

Microsoft Outlook, which is not to be confused (使困惑；把……混同) with Outlook Express, is personal information manager and e-mail communication software. The replacement for Microsoft Mail, Windows Messaging and Schedule starting in Office 97, it includes an e-mail client, calendar, task manager and address book.

Tell whether the following statements are True (T) or False (F).

(   ) 1. The used software in office automation only includes Microsoft, Corel and KingSoft.
(   ) 2. The first version of Office contained three desktop applications.
(   ) 3. Microsoft Excel was previously considered to be the main program in Office.
(   ) 4. Microsoft PowerPoint is used to create slideshows, composed of text, graphics, movies and other objects.
(   ) 5. Microsoft Outlook is a kind of personal information manager and e-mail communication software.

Choose one software to finish the task. And show some examples.

A. Word processing software   B. Web authoring software
C. E-mail software   D. Spreadsheet software
E. Presentation software   F. Database software

What you want to do	What kind of software you need	Example
Create text (letters, term papers, etc.)		
Set up a worksheet		
Build a presentation		
Build Web sites		
Communicate with other people		
Deal with a large amount of data, calculations		

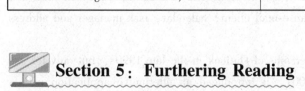

Section 5: Furthering Reading

Microsoft Office

Microsoft Office is a proprietary commercial Office suite of inter-related desktop applications, servers and services for the Microsoft Windows and Mac OS X operating systems, introduced by

Microsoft on August 1, 1989. Initially a marketing term for a bundled set of applications, the first version of Office contained Microsoft Word, Microsoft Excel, and Microsoft PowerPoint. Over the years, Office applications have grown substantially closer with shared features such as a common spell checker, OLE data integration and Microsoft Visual Basic for Applications scripting language. Microsoft also positions Office as a development platform for line-of-business software under the Office Business Applications brand.

The current versions are Office 2010 for Windows, released on June 15, 2010; and Office 2011 for Mac OS X, released on October 26, 2010.

Components

Microsoft Word is a word processor and was previously considered the main program in Office. Its proprietary DOC format is considered a de facto standard, although Word 2007 can also use a new XML-based, Microsoft Office-optimized format called DOCX, which has been controversially standardized by Ecma International as Office Open XML and its SP2 update supports PDF and a limited ODF. Word is also available in some editions of Microsoft works. It is available for the Windows and Mac platforms. The first version of Word, released in the autumn of 1983, was for the MS-DOS operating system and had the distinction of introducing the mouse to a broad population. Word 1.0 could be purchased with a bundled mouse, though none was required. Following the precedents of LisaWrite and MacWrite, Word for Macintosh attempted to add closer WYSIWYG features into its package. Word for Mac was released in 1985. Word for Mac was the first graphical version of Microsoft Word. Despite its bugginess, it became one of the most popular Mac applications.

Excel

Microsoft Excel is a spreadsheet program that originally competed with the dominant Lotus 1 − 2 − 3, but eventually outsold it. It is available for the Windows and Mac platforms. Microsoft released the first version of Excel for the Mac in 1985, and the first Windows version (numbered 2.05 to line up with the Mac and bundled with a standalone Windows run-time environment) in November, 1987.

Outlook/Entourage

Microsoft Outlook(not to be confused with Outlook Express) is a personal information manager and e-mail communication software. The replacement for Windows Messaging, Microsoft Mail and Schedule + starting in Office 97, it includes an e-mail client, calendar, task manager and address book.

On the Mac, Microsoft offered several versions of Outlook in the late 1990s, but only for use with Microsoft Exchange Server. In Office 2001, it introduced an alternative application with a slightly different feature set called Microsoft Entourage. It reintroduced Outlook in Office 2011, replacing Entourage.

PowerPoint

Microsoft PowerPoint is a popular presentation program for Windows and Mac. It is used to create slideshows, composed of text, graphics, movies and other objects, which can be displayed on-screen and navigated through by the presenter or printed out on transparencies or slides.

Common Features

Most versions of Microsoft Office (including Office 97 and later) use their own widget set and do not exactly match the native operating system. This is most apparent in Microsoft Office XP and 2003, where the standard menus were replaced with a colored flat looking, shadowed menu style. The user interface of a particular version of Microsoft Office often heavily influences a subsequent version of Microsoft Windows. For example, the toolbar, colored buttons and the gray-colored "3D" look of Office 4.3 were added to Windows 95. The Ribbon, introduced in Office 2007, has been incorporated into several applications bundled with Windows 7.

Users of Microsoft Office may access external data via connection-specifications saved in "Office Data Connection" files.

Both Windows and Office use Service Packs to update software. Office used to release non-cumulative Service Releases, which were discontinued after Office 2000 Service Release 1.

Programs in past versions of Office often contained substantial Easter eggs. For example, Excel 97 contained a reasonably functional flight-simulator. Versions starting with Office XP have not contained any Easter eggs in the name of Trustworthy Computing.

Answer the following questions according to the passage.

1. In Office 2011 for the Mac, Microsoft offers ____ as a personal information manager and e-mail communication software.
   A. Entourage   B. Exchange Server
   C. Outlook   D. Outlook Express

2. ____ is a popular presentation program for Windows and Mac.
   A. Word   B. Excel
   C. Outlook   D. PowerPoint

3. Following the precedents of LisaWrite and MacWrite, Word for Macintosh attempted to add closer ____ features into its package.
   A. WYGIWYS   B. WYCIWYG   C. WYSIWYC   D. WYSIWYG

4. Users of Microsoft Office may access external data via connection-specifications saved in ____ files.
   A. doc   B. docx   C. XML   D odc

Self-checklist

根据实际情况，从 A、B、C、D 中选择合适的答案：A 代表你能很好地完成该任务；B 代表你基本上可以完成该任务；C 代表你完成该任务有困难；D 代表你不能完成该任务。

A B C D

□ □ □ □ 1. 理解并正确朗读听说部分的句子，正确掌握发音和语调。

□ □ □ □ 2. 模仿听说部分的句型进行简单的对话。

☐ ☐ ☐ ☐ 3. 读懂本课的短文，并正确回答相关问题。
☐ ☐ ☐ ☐ 4. 向同学介绍不同的应用软件及其特征等相关的基本知识。
☐ ☐ ☐ ☐ 5. 掌握并能运用本单元所学重点句型、词汇和短语。
☐ ☐ ☐ ☐ 6. 使用电子词典查阅 Furthering Reading，了解常用软件的最新信息。

Unit 5

Multimedia

Unit Goals

In this unit, we are focusing on the following issues:
★ Introduction to multimedia
★ The categories of multimedia
★ The history of multimedia
★ Exercises about multimedia
★ The prospects for multimedia

 Warm-up

The pictures below show different content forms combined in multimedia. Write the correct letters into the brackets.

A   B   

C

D   E   F

1. (　　) Interactivity   2. (　　) Audio   3. (　　) Still Images
4. (　　) Text   5. (　　) Animation   6. (　　) Video

 Section 1: Dialogue

Directions: Read the following dialogue in pairs and talk about the new experience brought by IMAX.

Situation: *Simon and George are neighbors and they are sharing their experience of watching IMAX movies.*

The Ultimate Movie Experience with IMAX

Simon: Hi, George. Didn't see you last night, where have you been?
George: Hi, Simon! I've been to the cinema and saw an IMAX film with my girlfriend.
Simon: IMAX? So what's that?
George: IMAX is a motion picture film format. It is a set of proprietary cinema projection standards created by the Canadian company IMAX Corporation, and that is why it is called IMAX. IMAX has the capacity to record and display images of far greater size and resolution than conventional film systems.
Simon: Oh, so it must be quite different from the traditional movies.
George: Yes, IMAX increases the resolution of the image by using a much larger film frame. So the pictures are more lively. And what's more, IMAX theaters place speakers both directly behind the screen and around the theater to create a three-dimensional effect.
Simon: Wow! That must be fantastic!
George: Yeah, that's true! I strongly recommend you to go to an IMAX theater to have a try.
Simon: Good idea. Well, first of all, I have to ask Linda if she has time to go with me.

📧 Words & Expressions

proprietary [prəˈpraɪət(ə)rɪ]   *adj.* 所有的；专利的；私人拥有的
capacity [kəˈpæsɪtɪ]   *n.* 能力；容量
resolution [rezəˈluːʃ(ə)n]   *n.* 分辨率
conventional [kənˈvenʃ(ə)n(ə)l]   *adj.* 常见的，传统的
dimensional [dɪˈmenʃənəl]   *adj.* 空间的

🐫 Role Play

A: Excuse me, Miss/Mrs/Mr... What kinds of film do you often watch?
B: Well, it is/they are...

Unit 5 Multimedia

 Section 2: Reading

---

Pre-reading Activities

1. Have you heard of multimedia? Talk about the main purpose of the multimedia application.
2. What kinds of multimedia do you usually encounter in your daily lives?

Multimedia

Multimedia is media and content that uses a combination of different content forms. It is used in contrast to media which only use traditional forms of printed or hand-produced material. Multimedia includes a combination of text, audio, still images, animation, video, and interactivity content forms.

In the intervening forty years, the word has taken on different meanings. In the late 1970s the term was used to describe presentations consisting of multi-projector slide shows timed to an audio track. However, by the 1990s "multimedia" took on its current meaning.

Multimedia may be broadly divided into linear and non-linear categories. Linear active content progresses without any navigational control for the viewer such as a cinema presentation. Non-linear content offers user interactivity to control progress as used with a computer game or used in self-paced computer based training. Multimedia presentation can be live or recorded. A recorded presentation may allow interactivity via a navigation system. A live multimedia presentation may allow interactivity via an interaction with the presenter or performer.

Multimedia presentations may be viewed in person on stage, projected, transmitted, or played locally with a media player. A broadcast may be a live or recorded multimedia presentation. Broadcasts and recordings can be either analog or digital electronic media technology. Digital online multimedia may be downloaded or streamed. Streaming multimedia may be live or on-demand.

The various formats of technological or digital multimedia may be intended to enhance the users' experience, for example to make it easier and faster to convey information. Or in entertainment or art, to transcend everyday experience.

Enhanced levels of interactivity are made possible by combining multiple forms of media content. Online multimedia is increasingly becoming object-oriented and data-driven, enabling applications with collaborative end-user innovation and personalization on multiple forms of content over time. In addition to seeing and hearing, haptic technology enables virtual objects to be felt. Emerging technology involving illusions of taste and smell may also enhance the multimedia experience.

a live multimedia performance

📧 Words & Expressions

multimedia [ˈmʌltɪmiːdɪə]	n. 多媒体
combination [kɒmbɪˈneɪʃ(ə)n]	n. 组合，联合
linear [ˈlɪnɪə]	adj. 线性的
nonlinear [nɒnˈlɪnɪə]	adj. 非线性的
interactivity [ɪntərˈæktɪvɪtɪ]	n. 互动性
intervening [ˌɪntəˈviːnɪŋ]	adj. 中介的，介于其间的
presentation [prez(ə)nˈteɪʃ(ə)n]	n. 显示，表演
self-paced [selfˈpeɪst]	adj. 自定进程的，自定步调的
via [ˈvaɪə]	prep. 经由，通过
navigational [ˌnævɪˈgeɪʃnəl]	adj. 航海的，航行用的
navigation [ˌnævɪˈgeɪʃ(ə)n]	n. 航海；航空；航行
project [prəˈdʒekt]	v. 放映，投影
transmit [trænzˈmɪt]	v. 传送，传播
analog [ˈænəlɒg]	n. 类似物，相似物
digital [ˈdɪdʒɪt(ə)l]	adj. 数字的
stream [striːm]	n. 流
audio [ˈɔːdɪəʊ]	n. & adj. 音频（的）
animation [ænɪˈmeɪʃ(ə)n]	n. 动画
format [ˈfɔːmæt]	n. 版本，形式
enhance [ɪnˈhɑːns]	v. 提高；增加
transcend [trænˈsend]	v. 超出，超过
orient [ˈɔːrɪənt]	n. 东方
object-oriented [ˈɒbdʒektɔːrɪentɪd]	adj. 面向对象的；对象趋向的
collaborative [kəˈlæbərətɪv]	v. 协作，合作
innovation [ˌɪnəˈveɪʃn]	n. 革新，创新
haptic [ˈhæptɪk]	adj. 触觉的
virtual [ˈvɜːtʃʊəl]	adj. 事实上的，实际上的

emerge [ɪˈmɜːdʒ]   v. 浮现，出现

 Basic Technical Terms

still images	静态图像
multimedia presentation	多媒体演示
digital electronic media	数字式电子媒介技术
digital online multimedia	数字式网上多媒体

Choose the best answer according to the text.

(   ) 1. Multimedia is _____.
   A. a program
   B. a combination of different content forms
   C. an operating system

(   ) 2. Multimedia is _____.
   A. the same as common media   B. similar to traditional media
   C. different from traditional media

(   ) 3. Multimedia has the current meaning _____.
   A. since the beginning   B. in the late 1970s   C. by the 1990s

(   ) 4. Multimedia presentations can be _____.
   A. live   B. recorded   C. live or recorded

(   ) 5. A live multimedia presentation may _____.
   A. allow interactivity via an interaction with the presenter or performer
   B. allow interactivity via a navigation system
   C. not allow interactivity

(   ) 6. By adding _____ to your programs, you can make computers more interesting and much more fun for the user.
   A. multimedia   B. text   C. music

 Section 3：Computer Terms

Do You Know?

　　Multimedia：多媒体，是多种媒体的综合，一般包括文本、声音和图像等多种媒体形式。
　　在计算机系统中，多媒体指组合两种或两种以上媒体的一种人机交互式信息交流和传播媒体。使用的媒体包括文字、图片、照片、声音、动画和影片，以及程序所提供的互动功能。
　　多媒体是超媒体（Hypermedia）系统中的一个子集，而超媒体系统是使用超链接（Hyperlink）构成的全球信息系统，全球信息系统是因特网上使用 TCP/IP 协议和 UDP/IP 协

议的应用系统。二维的多媒体网页使用 HTML、XML 等语言编写，三维的多媒体网页使用 VRML 等语言编写。许多多媒体作品使用光盘发行，以后将更多地使用网络发行。

IMAX：Image Maximum 的缩写，被称为巨幕电影。它是一种能够放映比传统胶片更大和更高解像度的电影放映系统。整套系统包括以 IMAX 规格摄制的影片复制、放映机、音响系统、银幕等。标准的 IMAX 银幕为 22 米宽、16 米高，但完全可以在更大的银幕播放，而且迄今为止不断有更大的 IMAX 银幕出现。IMAX 的构造亦与普通电影院有很大分别。由于画面分辨率提高，观众可以更靠近银幕，一般所有座位均在一个银幕的高度内（传统影院座位跨度可达到 8～12 个银幕）；此外，座位倾斜度亦较大（在半球形银幕的放映室可倾斜达 23°），便于观众面向银幕中心。

MOOC：Massive Open Online Courses 的缩写，被称为大型开放式网络课程。2012 年，美国的顶尖大学陆续设立网络学习平台，在网上提供免费课程，Coursera、Udacity、edX 三大课程提供商的兴起，给更多学生提供了系统学习的可能。这三个大平台的课程全部针对高等教育，并且像真正的大学一样，有一套自己的学习和管理系统。再者，它们的课程都是免费的。因为 MOOC 有为数众多的学习者，以及可能有相当高的学生—教师比例，它常常使用客观、自动化的线上评量系统，像是随堂测验、考试等来促进大量回应和互动的教学设计。

CAI：计算机辅助教学（Computer Aided Instruction），是在计算机辅助下进行的各种教学活动，以对话方式与学生讨论教学内容、安排教学进程、进行教学训练的方法与技术。CAI 为学生提供一个良好的个人化学习环境。综合应用多媒体、超文本、人工智能和知识库等计算机技术，具有交互性、多样性、个别性、灵活性等特点，克服了传统教学方式单一、片面的缺点。它的使用能有效地缩短学习时间、提高教学质量和教学效率，实现最优化的教学目标。CAI 能够最大限度地缩短学生接受课程内容的时间，大幅度增强教学效果，最终使教学目标达到最优化。

Section 4：Exercises

Ⅰ. **Tell whether the following statements are True（T）or False（F）.**

(　　) 1. Multimedia only uses traditional forms of printed or hand-produced material.
(　　) 2. Multimedia always has the same meaning.
(　　) 3. Linear active content progresses with no navigational control for the viewer such as a cinema presentation.
(　　) 4. A recorded presentation may allow interactivity via an interaction with the presenter or performer.
(　　) 5. Broadcasts and recordings can be neither analog nor digital electronic media technology.
(　　) 6. Multimedia presentations can only be recorded.
(　　) 7. Enhancing levels of interactivity is not impossible.
(　　) 8. People cannot enhance their experience by the different formats of multimedia.

Unit 5　Multimedia

Ⅱ. **Complete the following sentences according to the text.**

1. Multimedia includes a combination of _____, _____, _____, _____, _____ and _____.
2. Multimedia may be broadly divided into _____ and _____ categories.
3. Multimedia presentations can be _____ and _____.
4. A recorded presentation may allow interactivity via _____ while a live multimedia presentation may allow interactivity via _____.
5. A broadcast may be _____.
6. Combining multiple forms of media content make it possible to _____.

Ⅲ. **Ability to explore.**

(　　) 1. Which of the following is multimedia player software?
　　　A. IE　　　　　　　　　　　　　B. ACDSee
　　　C. Windows Media player　　　　D. Outlook Express

(　　) 2. Which of the following can we use to create multimedia flash e-cards?
　　　A. Use Ready-made Free Online E-cards　　B. Use Free Online E-card Makers
　　　C. Use PowerPoint　　　　　　　　　　　D. Use Adobe Flash

Ⅳ. **Fill in the table below by giving the corresponding translation with the help of dictionaries or Internet.**

English	Chinese
Audio/Video Interleave (AVI)	
	一种程序接口
Joint Photographic Experts Group (JPEG)	
	运动图像专家组
National Television Standards Committee (NTSC)	
	逐行倒相制（电视）

 Section 5：Furthering Reading

Adobe Photoshop

Adobe Photoshop is a kind of photo retouching, image editing, and color painting software. Whether you are a novice or an expert in image editing, the Photoshop program offers you the tools you need to get professional-quality results.

Photoshop provides integrated tools for creating and outputting crisp, editable vector shapes and text. With the new tools, you can incorporate resolution-independent, vector-based graphics and type into pixel-based images to achieve an unparalleled range of design effects.

The new rectangle, rounded rectangle, ellipse, polygon, and line tools let you create a wide variety of vector-based shapes. These tools can be used to create shape layers. Like Adobe Illustrator, Photoshop provides pathfinder operations—Add, Subtract, Restrict, and Invert—for quickly combining basic vector shapes into complex shapes.

With Photoshop, you can easily combine crisp, resolution-independent type with pixel-based images, and then output sharp type edges with your image to produce high-quality results. What's more, Photoshop includes extensive new type formatting controls to help you produce the best-looking text possible, including the new type-warping that lets you twist and pull type to produce cool effects. Best of all, the type remains directly editable in the image no matter how you manipulate it.

Ⅰ. Translate the following phases into Chinese.

1. photo retouching
2. image editing
3. vector-based graphics
4. rounded rectangle
5. formatting controls

Ⅱ. Identify the following to be True (T) or False (F) according to the text.

1. Photoshop provides integrated tools for creating and outputting crisp, editable vector shapes and text. (   )
2. Photoshop doesn't include extensive new type formatting controls to help you produce the best-looking text possible. (   )

Self-checklist

根据实际情况，从 A、B、C、D 中选择合适的答案：A 代表你能很好地完成该任务；B 代表你基本上可以完成该任务；C 代表你完成该任务有困难；D 代表你不能完成该任务。

A	B	C	D	
☐	☐	☐	☐	1. 理解并正确朗读听说部分的句子，正确掌握发音和语调。
☐	☐	☐	☐	2. 模仿听说部分的句型进行简单的对话。
☐	☐	☐	☐	3. 读懂本课的短文，并正确回答相关问题。
☐	☐	☐	☐	4. 向同学介绍多媒体的基本知识。
☐	☐	☐	☐	5. 掌握并能运用本单元所学重点句型、词汇和短语。
☐	☐	☐	☐	6. 使用电子词典查阅 Furthering Reading，了解常用多媒体。

Unit 6

Computer Networks

> **Unit Goals**
>
> In this unit, we are focusing on the following issues:
> ★ Introduction to computer networks
> ★ The working methods of computer networks
> ★ Exercises about computer networks
> ★ The latest information about computer networks

 Warm-up

Ⅰ. Work in small groups and try to list as many words related to the Internet as possible, such as "TCP/IP, e-mail, www, etc."

- _____
- _____
- _____
- _____
- _____

Ⅱ. Look at your list of words related to the Internet. Discuss the following questions with your partner.

1. What do you use most often? Why?
2. How do you use it?

 Section 1: Dialogue

Directions: Read the following dialogue in pairs and talk about how computers communicate with each other.

Situation: *Simon wants to set up a company of e-commerce. George is a computer expert. Simon is consulting George on arranging the computer network in his company.*

Computer Network Is Fascinating

Simon: Hi, George. I am going to set up an e-commerce company, which requires many computers. I just wonder how these computers can communicate with each other and work together effectively.

George: Well, there are some predefined rules, protocols, standards and devices that are involved in the overall communication process.

Simon: Protocols? What do you mean?

George: There's no way to clearly explain this in a word and the most commonly used protocol is TCP/IP. Just like two people in a conversation must understand the same language. You can think of TCP/IP as this language and it must be properly installed and configured on every computer. At the same time, every computer must process a LAN card with an updated driver and precise configuration.

Simon: What do computer networks work for?

George: Networks can be used for a variety of purposes. Using a network, people can communicate efficiently and easily via e-mails, chat rooms, video telephone calls, and video conferencing. In a networked environment, each computer may access and use hardware resources on the network, such as printing a document on a shared network printer. Authorized users may access data and information stored in other computers on the network. Many networks provide access to data and information on shared storage devices. Users connected to a network may run application programs on remote computers.

Simon: Oh, that's really fascinating and what about the ways that networks are linked?

George: Networks link in several different ways and the safest one is a star network where each workstation has its own direct line to the server. If one link fails, which sometimes happens, it does not affect others. Even though it is expensive to install, a star network is normally preferred because of its safety.

Simon: That's it, star network, since security comes first for a new company. Thank you so much, George.

George: With pleasure. By the way, congratulations on your new business, Simon!

📧 Words & Expressions

arrange [əˈreɪndʒ]	*v.* 排列；整理
protocol [ˈprəʊtəkɒl]	*n.*（信息交换）协议
TCP/IP	传输控制协议/因特网互联协议
video conferencing	电视会议
authorized [ˈɔːθəraɪzd] **user**	授权用户
remote [rɪˈməʊt]	*adj.* 远程的；遥远的

Unit 6 Computer Networks

 Role Play

A: Excuse me, Miss/Mrs/Mr... Do you know anything about network?
B: Well, it is...

 Section 2: Reading

Pre-reading Activities

Do you often surf the Internet? What do you often do when you surf the Internet? Do you know how the Internet works?

What Is the Internet and How Does It Work?

The Internet is a global system of interconnected computer networks that use the standard Internet Protocol Suite (TCP/IP) to serve billions of users worldwide. It is a network of networks that consists of millions of private, public, academic, business, and government networks, of local to global scope, that are linked by a broad array of electronic and optical networking technologies. The Internet carries a vast range of information resources and services, such as the inter-linked hypertext documents of the World Wide Web (WWW) and the infrastructure to support electronic mail.

What do you need to do if you want to surf the Internet?

If you know the address, just carefully type it into the box on screen. If you don't know the address, go to a popular search engine like Google (*www.google.com*) or Baidu (*www.baidu.com*). Once there, you just need to type in some key words. For example, you may want to find out about the weather in your city. If you live in Nanjing, you would type in "Nanjing" and "weather." Then you get a list of sites. Choose the right site and you can have the information you want.

However, you are advised not to believe everything online. You have to watch out when you use information from the Internet. Anyone can create a website, so you cannot always be sure the information is correct. An official website produced by a big company or organization may be more useful than a site created by someone who isn't an expert. It is often best to use more than one source to make sure the information is correct.

 Words & Expressions

global [ˈɡləʊb(ə)l]   *adj.* 全球的；全世界的

— 47 —

interconnect [ˌɪntəkəˈnekt]	vi.	互相连接；互相联系
standard [ˈstændəd]	adj.	普通的；正常的；通常的；标准的
academic [ˌækəˈdemɪk]	adj.	学校的；学院的
scope [skəʊp]	n.	（处理、研究事务的）范围
array [əˈreɪ]	n.	展示；陈列；一系列
hypertext [ˈhaɪpətekst]	n.	超文本
infrastructure [ˈɪnfrəstrʌktʃə]	n.	基础设施；基础结构；基础建设
surf [sɜːf]	vi.	滑浪；（互联网上）冲浪；漫游
online [ɒnˈlaɪn]	adj.	（计算机）联机的；（与计算机）连线的
create [kriːˈeɪt]	vt.	创造；创作；创建；创设；设计
official [əˈfɪʃ(ə)l]	adj.	官方的；正式的；官方认可的
company [ˈkʌmp(ə)nɪ]	n.	公司
organization [ˌɔːgənaɪˈzeɪʃn]	n.	组织；团体；机构
expert [ˈekspɜːt]	n.	专家；能手；权威；行家；高手
source [sɔːs]	n.	来源；出处

📠 Basic Technical Terms

watch out	提防；小心
search engine	搜索引擎
World Wide Web	万维网

Complete the following sentences according to the passage.

1. The Internet is a _____ _____ of interconnected computer networks.
2. The Internet use the _____ _____ _____ _____ to serve billions of users worldwide.
3. You just need to type in some _____ _____ when you go to a popular search engine.
4. You _____ _____ everything online.
5. An _____ website produced by a big company or organization may be more useful than a site _____ _____ _____ who isn't an expert.

Section 3：Computer Terms

Do You Know?

　　Server：服务器。Server 一词，有软硬之分。从硬件的角度而言，Server 是物理上存在的服务器；而从软件的角度上说，Server 指的是具备服务器端功能的计算机软件，以及正在运行的服务器端软件。整个网络，是由无数的节点和连接通道共同构建而成的。从"硬"的方面说，是由无数的硬件服务器和其他数字化计算设备终端（比如个人计算机、手机等）

以及中间连接设备（比如网线、路由器等）构建而成的。从"软"的方面说，它是由无数运行着的服务器端软件和客户端软件（或者说终端软件）以及它们的相互连接交流而构建成的。无论"软"还是"硬"，无论是作为物理上的服务器还是服务器端的软件，Server 都始终存在着可靠性、高可用性和可扩充性的要求。

LAN：局域网（Local Area Network）。它是指在某一区域内由多台计算机互联而成的，一般是方圆几千米以内。局域网可以实现文件管理、应用软件共享、打印机共享、工作组内的日程安排、电子邮件和传真通信服务等功能。局域网是封闭型的，可以由办公室内的两台计算机组成，也可以由一个公司内的上千台计算机组成。决定局域网的主要技术要素为：网络拓扑、传输介质与介质访问控制方法。局域网由网络硬件（包括网络服务器、网络工作站、网络打印机、网卡、网络互联设备等）、网络传输介质，以及网络软件组成。

WAN：广域网（Wide Area Network）。它又被称为广域网、外网、公网，是连接不同地区局域网或城域网计算机通信的远程网，通常跨接很大的物理范围，所覆盖的范围从几十公里①到几千公里，它能连接多个地区、城市和国家，或横跨几个洲并能提供远距离通信，形成国际性的远程网络。广域网并不等同于互联网。广域网的发送介质主要是利用电话线或光纤，由 ISP 业者将企业间做连线，这些线是 ISP 业者预先埋在马路下的线路，因为工程浩大，维修不易，而且带宽是可以被保证的，所以成本会比较高。一般所指的互联网是属于一种公共型的广域网，公共型的广域网的成本较低，为一种较便宜的网上环境，但跟广域网相比较来说，它是没办法管理带宽的。走公共型网上系统的话，任何一段的带宽都无法被保证。

Ethernet：以太网。它是一种计算机局域网技术。IEEE 组织的 IEEE 802.3 标准制定了以太网的技术标准，它规定了包括物理层的连线、电子信号和介质访问层协议的内容。以太网是目前应用最普遍的局域网技术。它实现了网络上无线电系统多个节点发送信息的想法，每个节点必须获取电缆或者信道才能传送信息，有时也叫以太（Ether）。每一个节点有全球唯一的 48 位地址，也就是制造商分配给网卡的 MAC 地址，以保证以太网上所有节点能互相鉴别。由于以太网十分普遍，许多制造商把以太网卡直接集成到计算机主板上。以太网通信具有自相关性的特点，这对于电信通信工程十分重要。

Bus Network：总线型网络。它使用一定长度的电缆，也就是必要的高速通信链路将设备（比如计算机和打印机）连接在一起。设备可以在不影响系统中其他设备工作的情况下从总线中取下。总线型网络中最主要的就是以太网，它目前已经成为局域网的标准。这种结构具有费用低、数据端用户入网灵活、站点或某个端用户失效不影响其他站点或端用户通信的优点。缺点是一次仅能一个端用户发送数据，其他端用户必须等待到获得发送权。尽管有上述一些缺点，但由于布线要求简单、扩充容易、端用户失效、增删不影响全网工作，所以是 LAN 技术中使用最普遍的一种。

Ring Network：环型网络。它是使用一个连续的环将每台设备连接在一起。它能够保证一台设备上发送的信号可以被环上其他所有的设备都看到。在简单的环型网络中，网络中任何部件的损坏都将导致系统出现故障，这样将阻碍整个系统进行正常工作。具有高级结构的环型网络在很大程度上改善了这一缺陷。这种结构的网络型式主要应用于令牌网中，在这种

---

① 1 公里 = 1 000 米。

网络结构中各设备是直接通过电缆来串接的，最后形成一个闭环，整个网络发送的信息就是在这个环中传递，通常把这类网络称为"令牌环网"。

Star Network：星型网络。在星型（也称Star）拓扑中，网络中所有的计算机均连接至同一中枢装置（如交换机或集线器），每台计算机都分别通过一根电缆与该中枢装置相连接，集线器位于网络的中心位置，网络中的计算机都从这一中心点辐射出来，看上去就像是星星放射出的光芒，这或许就是当初称该种拓扑结构为星型的原因。现在，几乎所有的网络都采用星型拓扑，或者是由星型拓扑延伸出来的树型拓扑。由于星型网络中所有计算机都直接连接到集线设备（交换机或集线器）上，当一台计算机与另外一台计算机进行通信时，都必须经过中心节点。因此，可以在中央节点执行集中传输控制策略。集中传输控制使得网络的协调与管理更容易，网络带宽的升级更加简单，但也成为一个潜在的影响网络速度的"瓶颈"。

Section 4：Exercises

Ⅰ. **The pictures below are about computer. Please label the pictures and complete the following sentences with the given words and phrases.**

  E-mail _____   FTP _____
  WWW _____   Telnet _____

1. _____ is a method of transferring files from one computer to another over the

Unit 6 Computer Networks

Internet, even if each computer has a different operating system or storage format.
2. The _____, which Hypertext Transfer Protocol (HTTP) works with, is the fastest growing and most widely-used part of the Internet.
3. _____ allows an Internet user to connect to a distance computer and use that computer as if he or she were using directly.
4. The most widely used tool on the Internet is _____. It enables you to send messages to other areas, no matter how far between individuals.

Ⅱ. **Read the text and decide whether the following statements are True or False.**
(　　) 1. The Internet is a huge network, connecting computers to computers across the world.
(　　) 2. If you don't know the address, you could never get to the website you want.
(　　) 3. Search engine is a useful tool when you surf the Internet.
(　　) 4. Only big companies and governments are allowed to create a website.
(　　) 5. The Internet is powerful and everything in it is true.

Ⅲ. **Choose the best answer according to the passage.**
(　　) 1. The Internet connects _____ from around the world so people can share information.
　　　A. computers　　　B. telephones　　　C. radios　　　D. televisions
(　　) 2. Which of the following statements is NOT true?
　　　A. To get to the website you want, you need to use the correct address.
　　　B. If you don't know the address, go to a popular search engine.
　　　C. The World Wide Web is the part of the Internet that provides you with information in a form.
　　　D. You are advised not to believe everything online.
(　　) 3. What kind of website is more believable?
　　　A. A site created by someone.
　　　B. A site created by an expert.
　　　C. A site created by a teacher.
　　　D. A site created by a big company or organization.
(　　) 4. The word "Internet" describes the _____ that is believed to have got started during the 80's of the 20th century.
　　　A. computer　　　　　　　　　　B. picture
　　　C. network　　　　　　　　　　　D. work
(　　) 5. Which one of the following is not a search engine?
　　　A. Google.　　　　　　　　　　　B. Sougou.
　　　C. Baidu.　　　　　　　　　　　　D. Office.

Ⅳ. **Ability to explore.**
　　(　　) What may be the cause, if you can't visit a webpage?

  A. You typed the wrong address.
  B. Your computer are not connected to the Internet.
  C. There is something wrong with your computer.
  D. All of the above.

Ⅴ. **Fill in the table below by giving the corresponding translation with the help of dictionaries or Internet**.

English	Chinese
DNS (Domain Name Server)	
	文件传送协议
ISDN (Integrated Services Digital Network)	
	局域网
POP (Post Office Protocol)	
	对等协议
SLIP (Serial Line Interface Protocol)	
	简单邮件传送协议
TCP/IP (Transfer Control Protocol/Internet Protocol)	
	统一资源定位器

Section 5: Furthering Reading

Mobile Internet

  Is your smartphone or tablet always connected to the Internet? Millions have this always on connection to access their e-mail, favorite websites, cloud services, and APPs from anywhere at any time. What can be confusing to many is the variety of connection devices, data plans, and penalties for exceeding usage limits.

  There are many devices that can help you get Internet access from wherever you are:

  (1) Smartphones. Most modern smartphones are capable of accessing the Internet via WiFi and 3G or 4G networks.

  (2) Tablets. Most tablets will provide WiFi access, but to get 3G or 4G capabilities, you will generally have to purchase a higher-priced model.

  (3) Laptops. By inserting a USB modem, almost all laptops can access the Internet through a 3G or 4G network.

  Mobile hotspot device is a standalone device that connects to a 3G or 4G network. It will then allow multiple devices near it to access the Internet via a WiFi connection.

Many smartphones can provide WiFi sharing, or tethering, like a mobile hotspot device. However, you may have to pay an additional monthly fee.

Identify the letter of the choice that best matches the phrase or definition.

a. smartphone
b. tablet
c. laptop
d. WiFi
e. hotspot

_____ 1. Broadband Internet connection that uses radio signals to provide connections to computers and devices with build-in WiFi capability or a communications device that enables WiFi connectivity.

_____ 2. Thin, lightweight mobile computer with a screen in its lid and a keyboard in its base, designed to fit on your lap.

_____ 3. A wireless network that provides Internet connections to mobile computers devices.

_____ 4. An Internet-capable phone that usually also includes a calendar, an appoint book, an address book, a calculator, a notepad, game, browser, and numerous other APPs.

_____ 5. Thin, lighter weight mobile computer that has a touch screen, usually smaller than laptop but larger than a phone.

Self-checklist

根据实际情况，从 A、B、C、D 中选择合适的答案：A 代表你能很好地完成该任务；B 代表你基本上可以完成该任务；C 代表你完成该任务有困难；D 代表你不能完成该任务。

A B C D
☐ ☐ ☐ ☐ 1. 理解并正确朗读听说部分的句子，正确掌握发音和语调。
☐ ☐ ☐ ☐ 2. 模仿听说部分的句型进行简单的对话。
☐ ☐ ☐ ☐ 3. 读懂本课的短文，并正确回答相关问题。
☐ ☐ ☐ ☐ 4. 向同学介绍计算机网络的基本知识。
☐ ☐ ☐ ☐ 5. 掌握并能运用本单元所学重点句型、词汇和短语。
☐ ☐ ☐ ☐ 6. 使用电子词典查阅 Furthering Reading，了解网络的最新信息。

Unit 7

Computer Security

Unit Goals

In this unit, we are focusing on the following issues:
★ The conception of computer security
★ The technologies of computer security
★ The classifications of computer viruses
★ The measures to ensure the security of computers

 Warm-up

Match the following antivirus software with its Chinese meanings.

1. Kaspersky   a. "间谍波特变种" 木马
2. Kingsoft Anti-virus   b. 金山毒霸
3. Win 32. Troj. Spyboter c. 瑞星杀毒软件
4. Norton   d. 诺顿
5. Rising Antivirus   e. 熊猫卫士
6. Panda Antivirus Pro 2010 f. 卡巴斯基

 Section 1: Dialogue

Directions: Read the following dialogue in pairs and talk about the safety problems you came across relating to a computer.

Situation: *Simon had some problems with his computer and he dialed George, a computer expert, asking for some information about how to repair and what to do next.*

Safety Problems of a Computer

Simon: Hello, this is Simon speaking. Is that George?
George: This is George speaking. Simon, what's the matter?
Simon: George, this is an emergency, my computer is down and I've no idea what to do about

Unit 7   Computer Security

it. System Restore can't seem find Windows, either.

George: Can you speak at length about this situation?

Simon: OK. Last night, my computer was running fine. This morning, it won't boot. Right after it gets done with the bios, I just get a flashing cursor, which never progresses any further. Can't get to the boot options menu, can't get to safe mode, can't even get a damn Windows logo.

George: Prior to this, have you installed software or visited any illegal websites?

Simon: My first thought was that the pirated software I installed last night might have sneaked something nasty past my virus scanner or just screwed up some settings. Also, Windows installed an update before shutdown!

George: You should pay attention to privacy and security, and do not install any unknown software. Then did you test to repair in some way?

Simon: Input my installation disk in the drive and reboot with the thought that I could just roll everything back to the most recent restore point and be on my way. That works fine. However, when I get to the installation menu, my computer turns into blue screen. Most alarming!

George: Don't worry, try to restart it later, and if it still doesn't work, send it to me. I'm available these days.

Simon: OK, thanks anyway. I always fear that my computer might be attacked by a virus.

George: You can use anti-virus software to protect your computer.

Simon: But some hackers might work out my passwords and they may break into my system.

George: Oh, it happens rarely. But when it happens, it is a nightmare for all of us.

✉ Words & Expressions

emergency [ɪˈmɜːdʒ(ə)nsɪ]		n. 紧急情况,突发事件
System Restore		系统恢复
speak at length about		详细讲述
boot [buːt]		v. 启动
Bios(Basic Input Output System)		基本输入输出系统
a flashing [ˈflæʃɪŋ] **cursor** [ˈkɜːsə]		闪动的光标
safe mode		安全模式
pirate [ˈpaɪərət] **software**		盗版软件
virus [ˈvaɪrəs] **scanner** [ˈskænə]		病毒扫描软件
screw [skruː] **up**		弄糟
shutdown [ˈʃʌtdaʊn]		n. 关机
anti-virus software		杀毒软件
password [ˈpɑːswɜːd]		n. 密码, 口令

Role Play

A: Excuse me, Miss/Mrs/Mr... Do you know any way to protect the security of personal information online?

B: Well, it is...

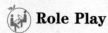
Section 2: Reading

Pre-reading Activity

Ⅰ. Below are some possible symptoms of a virus. Translate them into Chinese.

1. System becomes very slow.

2. A minute after starting the computer, it will turn itself off and then on again.

3. Your password is changed without notice.

4. You lose some file, but you did not delete.

5. Other people know your password or other private information.

6. A program disappears after you run it.

Ⅱ. Match the following viruses with their Chinese meanings.

1. Trojan horse   a. 玩笑和恶作剧
2. Backdoor   b. 黑客工具
3. Worm   c. 特洛伊木马
4. Joke   d. 蠕虫
5. Hacktool   e. 冲击波
6. Blaste   f. 后门程序

Computer Security

Computer security is a branch of computer technology known as Information Security as applied to computers and networks. The objective of computer security includes protection of information and property from theft, corruption, or natural disaster, while allowing the information and property to remain accessible and productive to its intended users.

Unit 7 Computer Security

The term computer system security means the collective processes and mechanisms by which sensitive and valuable information and services are protected from publication, tampering or collapse by unauthorized activities or untrustworthy individuals and unplanned events respectively. The strategies and methodologies of computer security often differ from most other computer technologies because of its somewhat elusive objective of preventing unwanted computer behavior instead of enabling wanted computer behavior.

The technologies of computer security are based on logic. As security is not necessarily the primary goal of most computer applications, designing a program with security in mind often imposes restrictions on that program's behavior.

A computer virus is a computer program that can copy itself and infect a computer. The term "computer virus" is sometimes used as a catch-all phrase to include all types of Malware, even those that do not have the reproductive ability. Malware includes computer viruses, computer worms, Trojan horses, most root kits, Spyware, dishonest Adware and other malicious and unwanted software, including true viruses.

Just as human viruses invade a living cell and then turn it into a factory for manufacturing viruses, computer viruses are small programs that replicate by attaching a copy of themselves to another program. Once attached to the host program, the virus then lock for other programs to "infect." In this way, the virus can spread quickly throughout a hard disk or an entire organization if it infects a LAN (Local Area Network) or a multi-users system. Viruses can increase their chances of spreading to other computers by infecting files on a network file system or a file system that is accessed by another computer.

In order to ensure the security of computers, we have to take some measures actively and consciously.

(1) Be very cautious about inserting disks from unknown sources into your computer.
(2) Always scan the disk's files before operating any of them.
(3) Only download Internet files from reputable sites.
(4) Do not open e-mail attachments (especially executable files) from strangers.

There is a lot of anti-virus software and other preventive measures. Many users install anti-virus software that can detect and eliminate known viruses after the computer downloads or runs the executable. This practice is known as "on-access scanning." Anti-virus software does not change the underlying capability of host software to transmit viruses. Users must update their software regularly to patch security holes. Anti-virus software also needs to be regularly updated in order to recognize the latest threats.

📧 Words & Expressions

strategy [ˈstrætədʒɪ]	n. 策略；战略；战略学
objective [ˌəbˈdʒektɪv]	n. 方法论；方法学
property [ˈprɒpətɪ]	n. 特性；属性；财产
theft [θeft]	v. 偷盗；偷窃；被盗
corruption [kəˈrʌpʃ(ə)n]	n. 腐败；贪污；贿赂
accessible [əkˈsesɪb(ə)l]	adj. 易接近的；可理解的；易相处的
collective [kəˈlektɪv]	adj. 集体的；共同的；集合的
mechanism [ˈmek(ə)nɪz(ə)m]	n. 机能；机制；结构
tamper [ˈtæmpə]	v. 篡改；窜改
collapse [kəˈlæps]	v. 崩溃；倒塌；折叠
methodology [meθəˈdɒlədʒɪ]	n. 方法论；方法学
elusive [ɪˈl(j)uːsɪv]	adj. 难以捉摸的；逃避的
restriction [rɪˈstrɪkʃ(ə)n]	n. 限制；限定；拘束；束缚；管制
infect [ɪnˈfekt]	v. 感染；传染；散布病毒；侵染
malicious [məˈlɪʃəs]	adj. 生殖的；再生产的；复制的
replicate [ˈreplɪkeɪt]	v. 复制；复写；重复；反复
preventive [prɪˈventɪv]	adj. & v. 预防性的，防备的；预防措施
executable [ɪgˈzekjʊtəb(ə)l]	adj.（可）执行的；实行的
underlying [ʌndəˈlaɪɪŋ]	adj. 潜在的；含蓄的；基础的
threat [θret]	n. & v. 威胁；恐吓；凶兆

🖥 Basic Technical Terms

computer security	计算机安全
computer virus	计算机病毒
computer technology	计算机技术
information security	信息安全
intended users	预期用户
the collective processes	系列过程
untrustworthy individual	不可信任的个体
dishonest adware	不可信任的恶意广告
computer application	计算机应用
the reproductive ability	复制能力
a network file system	网络文件系统
a multi-user system	多用户系统
preventive measure	预防措施

Unit 7　Computer Security

underlying capability　　　　　　　　潜在能力

Mark the following sentences with true or false according to the passage.

(　　) 1. Computer security is a branch of computer technology known as Information Security as applied to computers and networks.
(　　) 2. There are various strategies and techniques used to design security systems.
(　　) 3. Worms and Trojan horses are the same thing technically.
(　　) 4. Be very cautious about inserting disks from unknown sources into your computer.
(　　) 5. As new viruses are created every day, upgrade your anti-virus software regularly.

 Section 3：Computer Terms

Do You Know?

Internet Security：互联网安全，是一门涉及计算机科学、网络技术、通信技术、密码技术、信息安全技术、应用数学、数论、信息论等多种学科的综合性学科。互联网安全从其本质上来讲就是互联网上的信息安全。从广义上来说，凡是涉及互联网上信息的保密性、完整性、可用性、真实性和可控性的相关技术和理论都是网络安全的研究领域。

Virus：计算机病毒，复数是 viruses，通常是由来路不明的文档带入，计算机受到病毒感染，就像人体遭受细菌侵入一样，会因不同的病毒或细菌感染而产生不同的症状。目前虽有许多防毒软件，但病毒的产生是日新月异、不断增加的。因此最好的防毒方法就是尽量少去不熟悉的网站下载软件；另外，如果陌生人寄来的电子邮件中附有文档，最好不要任意开启！

Firewall：防火墙，也称防护墙，是由 Check Point 创立者 Gil Shwed 于 1993 年发明并引入国际互联网的［US5606668（A）1993-12-15］。防火墙是位于内部网和外部网之间的屏障，它按照系统管理员预先定义好的规则来控制数据包的进出。防火墙是系统的第一道防线，其作用是防止非法用户的进入。

Hacker：黑客，通常是指对计算机科学、编程和设计方面具高度理解的人。在信息安全里，"黑客"指研究智取计算机安全系统的人员。利用公共通信网络，如互联网和电话系统，在未经许可的情况下，载入对方系统的被称为黑帽黑客（英文：Black Hat；另称Cracker）；调试和分析计算机安全系统的被称为白帽黑客（英语：White Hat）。"黑客"一词最早用来称呼研究盗用电话系统的人士。

Pirated Software：盗版软件，是指非法制造或复制的软件。它非常难以识别，但缺少密钥代码或组件是缺乏真实性的表现。盗版软件无安全保证，无法提供合法的版权证书，不能为用户提供售后服务，并且无法升级。另外，还可能无法通过 Internet 下载软件供应商提供的更新。

Intellectual Property Rights：知识产权，是指"权利人对其所创作的智力劳动成果所享有的专有权利"，一般只在有限时间期内有效。各种智力创造比如发明、文学和艺术作品，以

及在商业中使用的标志、名称、图像以及外观设计，都可被认为是某一个人或组织所拥有的知识产权。主要分类有：专利权，商标权，著作权（版权）等。

 Section 4：Exercises

Ⅰ. **Match the items listed in the following two columns.**

1. virus detection　　　　　a. 系统还原
2. floppy disk　　　　　　 b. 软盘
3. Trojan horses　　　　　 c. 病毒检测
4. system restore　　　　　d. 木马
5. registration number　　　e. 注册码
6. file system　　　　　　 f. 文件系统
7. previous checkpoint　　　g. 还原点
8. boot sector　　　　　　 h. 引导扇区

Ⅱ. **Which are the potential possibilities if your computer breaks down?**

　　A. Unknown hard disk error.　　　　B. Virus attacks.
　　C. Insufficient memory.　　　　　　D. Too many open files.

Ⅲ. **Translation.**

1. Computer security is a branch of computer technology known _____（信息安全中的计算机和网络领域）.

2. As security is not necessarily the primary goal of most computer applications, designing a program with security in mind often _____（对程序的运行有所限制）.

3. The term "computer virus" is sometimes used as a catch-all phrase to include all types of Malware, _____（甚至是那些不具有复制能力的恶意软件）.

4. Many users install anti-virus software _____（可以侦测和消除已知病毒）after the computer downloads or runs the executable.

5. Computer viruses are small programs that _____（通过将该程序的本身附加到另一个程序上来进行复制）.

6. Anti-virus software also needs _____（定期更新，以识别最新的威胁）.

Ⅳ. **Reading comprehension.**

　　Until quite recently the opportunities for criminal activity on the Internet have been low. However, the volume of business done on the Internet is growing rapidly, as people order books

Unit 7　Computer Security

and other products and make money transactions（交易）. All this is creating temptations（诱惑）for hackers.

Hackers are often young people who are obsessed（使着迷）by computers. They use them to prowl（潜行于）the Internet, looking for ways to break into computer systems run by banks, telephone companies and even government departments. They look for examples of credit cards and try to steal the numbers.

Recently in America, hackers have been caught testing the security system at the Pentagon（五角大楼）, headquarters of the American Defense Department. But still the hackers persist for a dare "because it's there" although with what success nobody really knows.

The two ways to defeat hackers are:

1) To have an up-to-date virus scanner which can recognize the invaders and delete it.

2) To change one's password constantly.

Choose the best answer.

(　) 1. What is enemy number one on the Internet?
　　A. Virus.　　　　B. Hacker.　　　　C. Password.　　　　D. Crime.

(　) 2. What are hackers trying to steal?
　　A. Books.　　　　　　　　　　　　　B. Credit cards.
　　C. Security systems.　　　　　　　　D. Credit card numbers.

(　) 3. One of the effective ways against hackers is _____.
　　A. always changing computers　　　　B. always changing passwords
　　C. always changing systems　　　　　D. always changing alarms

(　) 4. In the passage hackers are compared to _____.
　　A. thieves　　　　　　　　　　　　　B. robbers
　　C. pickpockets　　　　　　　　　　　D. shoplifters

 ## Section 5: Furthering Reading

Passwords Are Everywhere in Computer Security

Passwords are everywhere in computer security. All too often, they are also ineffective. A good password has to be both easy to remember and hard to guess, but in practice people seem to pay attention to the former. Names of wives, husbands and children are popular. "123456" or "12345" are also common choices.

That predictability lets security researchers (and hackers) create dictionaries which list common passwords, useful to those seeking to break in. But although researchers know that passwords are insecure, working out just how insecure has been difficult. Many studies have only small samples to work on.

— 61 —

However, with the co-operation of Yahoo, Joseph Bonneau of Cambridge University obtained the biggest sample to date— 70 million passwords that came with useful data about their owners.

Mr Bonneau found some interesting variations. Older users had better passwords than young ones. People whose preferred language was Korean or German chose the most secure passwords; those who spoke Indonesian the least. Passwords designed to hide sensitive information such as credit-card numbers were only slightly more secure than those protecting less important things, like access to games. "Nag screens" that told users they had chosen a weak password made virtually no difference. And users whose accounts had been hacked in the past did not make more secure choices than those who had never been hacked.

But it is the broader analysis of the sample that is of most interest to security researchers. For despite their differences, the 70 million users were still predictable enough that a generic password dictionary was effective against both the entire sample and any slice of it. Mr Bonneau is blunt: "An attacker who can manage ten guesses per account will compromise around 1% of accounts." And that is a worthwhile outcome for a hacker.

One obvious solution would be for sites to limit the number of guesses that can be made before access is blocked. Yet whereas the biggest sites, such as Google and Microsoft, do take such measures, many do not. The reasons for their not doing so are various. So it's time for users to consider the alternatives to traditional passwords.

Answer the following questions according to the passage.

1. People tend to use passwords that are _____.
   A. easy to remember   B. hard to figure out
   C. random numbers   D. popular names
2. Researchers find it difficult to know how unsafe passwords are due to _____.
   A. lack of research tools   B. lack of research funds
   C. limited time of studies   D. limited size of samples
3. It is indicated in the text that _____.
   A. Indonesians are sensitive to password security
   B. young people tend to have secure passwords
   C. nag screens help little in password security
   D. passwords for credit cards are usually safe
4. The last paragraph of the text suggests that _____.
   A. net users regulate their online behaviors
   B. net users rely on themselves for security
   C. big websites limit the number of guesses
   D. big websites offer users convenient access

Self-checklist

根据实际情况，从A、B、C、D中选择合适的答案：A代表你能很好地完成该任务；

Unit 7　Computer Security

B 代表你基本上可以完成该任务；C 代表你完成该任务有困难；D 代表你不能完成该任务。

A　B　C　D

☐ ☐ ☐ ☐　1. 理解并正确朗读听说部分的句子，正确掌握发音和语调。

☐ ☐ ☐ ☐　2. 模仿听说部分的句型进行简单的对话。

☐ ☐ ☐ ☐　3. 读懂本课的短文，并正确回答相关问题。

☐ ☐ ☐ ☐　4. 向同学介绍计算机网络安全的基本知识。

☐ ☐ ☐ ☐　5. 掌握并能运用本单元所学重点句型、词汇和短语。

☐ ☐ ☐ ☐　6. 使用电子词典查阅 Furthering Reading，进一步了解网络安全。

Unit 8

E-commerce

> **Unit Goals**
>
> **In this unit, we are focusing on the following issues:**
> ★ What is e-commerce
> ★ The development of e-commerce
> ★ The categories of e-commerce
> ★ The advantages and disadvantages of e-commerce
> ★ The latest information about e-commerce

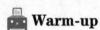 **Warm-up**

Look at the following pictures. There are some web sites from which people can buy things. Can you point them out from the list below?

The shopping sites are _____, _____, _____, _____, _____, _____

Section 1: Dialogue

Directions: Read the following dialogue in pairs and talk about your experience of shopping online.
Situation: *Simon and George are talking about what e-commerce is and exchanging their ideas about shopping online.*

Online Shopping and Some APPs

Simon: Do you know what e-commerce is, George?

George: It means we can buy things on the Internet.

Simon: Buy things online? Is that reliable? Have you bought anything through the net?

George: Of course, anything that you can imagine can be sought on some shopping platforms, like Alibaba, the biggest online shopping center in China. It is very convenient and fast and it is regarded as a shopper's heaven.

Simon: Then who sold things online?

George: Anyone who registered and was verified on the specific website can run the business.

Simon: Like me?

George: Definitely. As long as you find relevant supply of goods.

Simon: I would like to buy some books through the Internet? Can you recommend me some online book stores?

George: Dangdang is a good choice. I often buy books through Dangdang. You can try it! By the way, in order to facilitate online shopping experience, large suppliers like Taobao and Dangdang had their own APPs, which you can download to your mobile phone almost instantly.

Simon: How long shall I wait before I received the books?

George: It depends on the distance between the warehouse and the place of receipt, and you can check the express delivery information at any time.

Simon: That's amazing! So are there any disadvantages of online shopping?

George: Based on my shopping experience online so far, I should say, some fragile products may get damaged on the way if not properly packaged or put, the color and the textile of some clothes may fail to match with the description online sometimes and things like that.

Simon: Then how to avoid these unhappy experiences?

George: We should check the credits of the sellers and comments made by prior buyers.

Simon: Thanks a lot!

✉ Words & Expressions

platform [ˈplætfɔːm]	n.	平台
facilitate [fəˈsɪlɪteɪt]	v.	促进；帮助
supplier [səˈplaɪə]	n.	供应商
warehouse [ˈweəhaʊs]	n.	仓库
receipt [rɪˈsiːt]	n.	收据；收入
fragile [ˈfrædʒaɪl]	adj.	易碎的
textile [ˈtekstaɪl]	n.	纺织品
prior [ˈpraɪə]	adj.	在前的

 **Role Play**

A: Excuse me, Miss/Mrs/Mr... Do you often buy things online?
B: Yes, I usually...

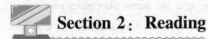 **Section 2: Reading**

Pre-reading Activities

I. Try to match the expressions with their Chinese meanings.

1. dot-com companies   a. 电子商务
2. electronic money   b. 网络公司
3. mobile commerce   c. 网上购物
4. electronic commerce   d. 移动商务
5. online shopping   e. 电子货币

II. Have you heard of the following terms of e-commerce? Try to match the activities with their definitions and see if you are right after reading the passage.

(1) Business-to-Consumer (B2C) E-commerce   a. occurs when a person sells products and services to another person through e-commerce.

(2) Business-to-Business (B2B) E-commerce   b. occurs when a company sells products and services through e-commerce to another company.

(3) Consumer-to-Consumer (C2C) E-commerce   c. occurs when a company sells products and services through e-commerce to customers who are individuals.

E-commerce

E-commerce, also known as e-business, is short for "electronic commerce." So what's "electronic commerce," then?

The meaning of e-commerce has changed over the last 30 years. At first, e-commerce meant the facilitation of commercial transactions electronically, using technology such as electronic funds transfer (EFT). Later, the growth of credit cards, automated teller machines (ATM) and telephone banking were also forms of e-commerce. Now we say e-commerce is the activity of buying or selling of products on online services or over the Internet. It draws on a variety of technologies such as mobile commerce, electronic funds transfer, supply chain management, Internet marketing, online transaction processing, electronic data interchange (EDI), inventory management systems, and automated data collection systems.

Some researchers categorize e-commerce by the types of entities participating in the transactions or business processes. The five general categories are business-to-consumer, business-to-business, consumer-to-consumer, business-to-government and consumer to government. And the three categories that are most commonly used are:

Consumer shopping on the Web, often called business-to-consumer (or B2C).

Transactions between businesses on the Web, often called business-to-business (or B2B).

Transactions and business processes in which companies, governments, and other organizations use Internet technologies to support selling and purchasing activities.

With the rapid development of Internet, the global e-commerce transaction has increased greatly within the past decades. Almost all kinds of industries are closely connected with e-commerce. The online market is expected to grow by 56% in 2015 – 2020. In 2017, retail e-commerce sales worldwide amounted to 2.3 trillion US dollars, which was a 25% increase than the previous year. And e-retail revenues are projected to grow to 4.88 trillion US dollars in 2021. Traditional markets are only expected 2% growth during the same time.

Among emerging economies, e-commerce presence in China continues to expand every year. With 668 million Internet users (twice as many as in the US), China is the world's biggest online market. China's online shopping sales reached $253 billion in the first half of 2015, accounting for 10% of the total Chinese consumer retail sales in that period. And it reached $899 billion in 2016.

E-commerce markets are growing at noticeable rates. However, everything has two sides. On one hand, booming e-commerce is the fastest way so far to make transactions across far distance. E-commerce makes it possible to do business at home, which saves time and unnecessary formalities. That's why e-commerce is preferable to traditional commerce. On the other hand, there exist many problems either. It is hard to control the virtual business. Relevant laws and regulations are imperfect. False, deceptive information is growing among e-commerces. Without management, losses are likely to happen every time. So everyone should hold strong risk awareness to protect themselves on e-commerce.

Words & Expressions

facilitation [fəsɪlɪˈteɪʃn]	n.	简易化
commercial [kəˈmɜːʃ(ə)l]	adj.	商业的，营利的
transaction [trænˈzækʃ(ə)n]	n.	交易，事务
interchange [ˈɪntətʃendʒ]	n.	互换
fund [fʌnd]	n.	资金；基金
transfer [trænsˈfɜ]	vt.	传递；转移
categorize [ˈkætəgəraɪz]	vi.	分类

entity [ˈentɪtɪ]	n. 实体；存在；本质
participate [pɑːˈtɪsɪpeɪt]	vi. 参与，参加
consumer [kənˈsjuːmə]	n. 消费者
purchase [ˈpɜːtʃəs]	vt. 购买；赢得
decade [ˈdekeɪd]	n. 十年，十年期
retail [ˈriːteɪl]	n. 零售
previous [ˈpriːvɪəs]	adj. 以前的，早先的
revenue [ˈrevənjuː]	n. 税收收入；财政收入；收益
trillion [ˈtrɪljən]	n. （数）万亿
emerge [ɪˈmɜːdʒ]	vi. 浮现，出现，显现
economy [ɪˈkɒnəmɪ]	n. 经济；经济体
expand [ɪkˈspænd]	vi. 扩张；发展；膨胀
noticeable [ˈnəʊtɪsəb(ə)l]	adj. 显而易见的；显著的；值得注意的
booming [ˈbuːmɪŋ]	adj. 兴旺的；繁荣的
formality [fɔːˈmælətɪ]	n. 礼节；仪式；正式手续
preferable [ˈpref(ə)rəb(ə)l]	adj. 更好的，更可取的，更合意的
virtual [ˈvɜːtʃʊəl]	adj. 虚拟的
relevant [ˈrelәvәnt]	adj. 相关的
regulation [regjʊˈleɪʃ(ә)n]	n. 管理；规则
deceptive [dɪˈseptɪv]	adj. 欺诈的；虚伪的
awareness [әˈweәnәs]	n. 意识，认识

Basic Technical Terms

e-commerce	电子商务
electronic funds transfer (EFT)	电子资金转移
automated teller machine (ATM)	自动柜员机，ATM 机
telephone banking	电话银行
mobile commerce	移动商务
supply chain management	供应链管理
Internet marketing	互联网营销
online transaction processing	在线交易处理
electronic data interchange (EDI)	电子数据交换
inventory management system	库存管理系统
automated data collection system	自动数据收集系统
business-to-consumer (B2C)	企业对消费者
business-to-business (B2B)	企业对企业
consumer-to-consumer (C2C)	消费者对消费者
business-to-government (B2G) 或 (B2A)	企业对政府
consumer to government (C2G) 或 (C2A)	消费者对政府

Unit 8　E-commerce

retail e-commerce sales　　　　　　零售电子商务销售额
e-retail revenues　　　　　　　　　电子零售收入
emerging economies　　　　　　　　新兴经济体
online shopping sales　　　　　　　在线销售额
total Chinese consumer retail sales　　中国零售总额

Answer these questions for comprehension.

1. What's e-commerce short for?

2. What's e-commerce?

3. What technologies does modern e-commerce mostly rely on?

4. What are the five general categories of e-commerce?

5. How much were the retail e-commerce sales worldwide in 2017?

 Section 3：Computer Terms

Do You Know?

　　Electronic Commerce：电子商务，从通信的角度解释，是利用电话线、网际网络等方式，传递资讯、产品、服务或付款服务；从企业流程的角度来看，是线上商业交易与工作流程自动化的电脑技术的应用。一般而言，电子商务的内容包括：资讯流、资金流、商流与物流。在网络新经济时代，电子商务（Electronic Commerce）成为企业新兴的营运模式，通过网际网络的网网相连，企业建构电子平台，进行交易与服务。

　　Electronic Money：电子货币，是指用一定金额的现金或存款从发行者处兑换并获得代表相同金额的数据或者通过银行及第三方推出的快捷支付服务，通过使用某些电子化途径将银行中的余额转移，从而能够进行交易。从严格意义上说，它是消费者向电子货币的发行者使用银行的网络银行服务进行储值和快捷支付，通过媒介（二维码或硬件设备），以电子形式使消费者进行交易的货币。

　　B2C：Business-to-Customer 的缩写，其中文简称为"商对客"。"商对客"是电子商务的一种模式，也就是通常说的直接面向消费者销售产品和服务的商业零售模式。

　　B2B：Business-to-Business 的缩写，其中文简称为"商对商"。它是指企业与企业之间通过专用网络或 Internet，进行数据信息的交换、传递，开展交易活动的商业模式。它将企业内部网和企业的产品及服务，通过 B2B 网站或移动客户端与客户紧密结合起来，通过网络的快速反应，为客户提供更好的服务，从而促进企业的业务发展。

　　C2C：Consumer-to-Consumer 的缩写，其中文是指个人与个人之间的电子商务。C2C 的意思就是消费者个人间的电子商务行为。比如一个消费者有一台计算机，通过网络进行交易，把它出售给另外一个消费者，此种交易类型就称为C2C电子商务。

B2A：Business-to-Administrations 的缩写，其中文是指商业机构对行政机构。它是企业与政府机构之间进行的电子商务活动。例如，政府将采购的细节在国际互联网上公布，通过网上竞价方式进行招标，企业也要通过电子的方式进行投标。

Section 4：Exercises

Ⅰ. Fill in the following blanks according to the passage.

1. E-commerce is the activity of buying or selling of products on _____ or over the _____.
2. Some researchers categorize e-commerce by the _____ participating in the transactions or business processes.
3. _____ e-commerce is usually referred to as "B2B."
4. _____ shopping on the Web, often called business-to-consumer (or B2C).
5. With the rapid development of _____, global e-commerce transaction has increased greatly.
6. Among emerging economies, _____ continues to expand every year.
7. On one hand, booming e-commerce is the fastest way so far to make transactions _____.
8. Without management, _____ are likely to happen every time.

Ⅱ. Complete the following sentences.

1. EFT _____ (是电子资金转移的缩写).
2. Later, _____ (信用卡、自动柜员机和电话银行的发展) were also forms of e-commerce.
3. Almost all kinds of industries are _____ (和电子商务密切相关).
4. In 2017, _____ (全球零售电子商务销售额) amounted to 2.3 trillion US dollars.
5. With 668 million Internet users, China is _____ (世界上最大的在线市场).
6. E-commerce markets are growing _____ (以引人注目的速度).

Ⅲ. Translate the following sentences into Chinese.

1. With the introduction of the Internet, companies, regardless of size, can communicate with each other electronically and cheaply.
2. It takes ten times more effort and more money to attract a new customer than to keep an existing customer.

3. Online shopping is becoming more and more popular in China.

4. E-bay is the largest Internet auction site in the world.

5. From the customer's point of view, shipping costs are a major concern.

Ⅳ. **Fill in the blanks with the correct words.**

Businesses can g_____ （获得） many benefits from engaging in electronic commerce. E-commerce can _____ （减少） the cost of purchases, l_____ （降低） sales and marketing costs and i_____ （改善） the c_____ （客户） service. However, businesses are not the only beneficiaries of e-commerce. Consumers may also r____ （收获） benefits. They can receive increased choice of vendors and p_____ （产品）, c_____ （便利） from shopping at home or office, greater amounts of i_____ （信息） that can be accessed on demand and more c_____ （有竞争力的） prices.

Ⅴ. **Ability to explore.**

How to Build Your Website of Your Own Web Shop

The best thing to do is to make a complete list of all the functionalities you want your website to have. This list must be comprehensive. Then, break that main list down into three smaller lists:

1. Your must-have list: This is a list of the functional elements that your site cannot go without.

2. Your nice-to-have list: This is a list of the functionalities that are not critical, but would really add to the user experience.

3. Your in-my-dreams list: This is a list of all the things that you feel are too time-, finance- or labor-intensive to do, but would be super cool.

Then, on a sheet of paper, do a mock-up drawing of how you want the site to be layed out. What you're looking for here is placement of various elements, not design. In professional terms, what you're doing is creating a first draft of your wire frames (you'll have to get these professionally redone later, but doing it this way for now is okay).

Only after you've done these steps should you try to find someone to help you build the site. It will also be helpful to determine whether you're going to pay someone or if you want to share equity or profits.

Then the next logical place is to start asking all the technical people you know for referrals. It's best to work from referrals. Be sure to interview at least three possible consultants so that you make sure you find the best person to help you.

How can you build your website if you want to have your own web shop? Put the following steps in a correct order.

(   ) A. Try to find someone to help you build the site.

( ) B. Ask all the technical people you know for referrals.
( ) C. Make a list of all the functionalities your website will have.
( ) D. Make a model drawing of how the site is to be layed out.
( ) E. Put the main list into three smaller lists.

Ⅵ. **Fill in the table below by giving the corresponding translation with the help of dictionaries or Internet.**

English	Chinese
logistics service	
	商业模式
small and medium-sized firms	
	银行账户
cancel the deal	
	原材料
second-hand market	
	降低成本
shipping cost	
	最低价
electronic payment	
	拍卖网站

Section 5: Furthering Reading

Dangdang Online Bookstore

Dangdang.com was established in November, 1999. Now it is regarded as the largest Chinese online bookstore. Dangdang.com provides more than 200,000 kinds of Chinese books and more than 10,000 kinds of audio and video products. Dangdang has set its position to become an online bookstore with "more choice, lower price" in the eyes of consumers.

Dangdang online bookstore has the following features.

(1) More types of goods. Dangdang has more than 200,000 kinds of books (accounts for 90% of the total books in Mainland China), more than 10,000 kinds of CD/VCD/DVD audio and video products as well as numerous games, software, EVDO, etc. It is an online retail store with the most varieties of goods.

(2) More convenient shopping. Dangdang has adopted an advanced method to classify goods. It

has intelligent Query and a simple shopping process, which provides consumers with a relaxing and pleasant shopping environment.

(3) A better distribution system. Dangdang has six large logistics centers and promotes COD service over 800 cities. It reflects the "customer-centered" principle, and manages to meet customers' needs better.

(4) A variety of promotion methods. The methods include establishment of group purchase, discount strategy on all the books, setup of sale markets with different discounts on targeted books, etc.

(5) "Internet Smart Parity" system. With the system, Dangdang monitors the online books and audio and video products information instantly. Whenever there is lower price on another website, it would automatically adjust to the same low price to maintain the competitive price advantage.

Fill the blanks with the words or phrases of the text.

1. Dangdang. com is regarded as the largest Chinese _____ _____.
2. Dangdang has become an online bookstore with "more _____, _____ price" in the eyes of consumers.
3. Dangdang promotes COD service, which reflects the _____ principle.
4. Dangdang. com ____ the online books and audio and video products information instantly to ____ to the same low price to maintain the competitive price _____.

Self-checklist

根据实际情况，从 A、B、C、D 中选择合适的答案：A 代表你能很好地完成该任务；B 代表你基本上可以完成该任务；C 代表你完成该任务有困难；D 代表你不能完成该任务。

A	B	C	D	
□	□	□	□	1. 理解并正确朗读听说部分的句子，正确掌握发音和语调。
□	□	□	□	2. 模仿对话部分的句型进行简单的对话。
□	□	□	□	3. 读懂本课的短文，并正确回答相关问题。
□	□	□	□	4. 向同学介绍电子商务的基本知识。
□	□	□	□	5. 掌握并能运用本单元所学重点句型、词汇和短语。
□	□	□	□	6. 使用电子词典查阅 Furthering Reading，进一步了解电子商务。

Unit 9

New Technologies

Unit Goals

In this unit, we are focusing on the following issues:
★ Introduction to Artificial Intelligence
★ The challenges in Artificial Intelligence technology
★ Exercises about Artificial Intelligence
★ The prospects for Artificial Intelligence

 Warm-up

Match the new technologies with their English terms on the right.

1. 量子计算　　　　　A. Block Chain
2. 人工智能　　　　　B. Cloud Computing
3. 云计算　　　　　　C. Internet of Things
4. 物联网　　　　　　D. Quantum Computation
5. 区块链技术　　　　E. Virtual Reality
6. 虚拟现实技术　　　F. Artificial Intelligence

 Section 1: Dialogue

Directions: Read the following dialogue in pairs and talk about your understanding of cloud computing.

Situation: *Simon and George are talking about what "Cloud Computing" is.*

Cloud Computing

Simon: Hello George! There is a popular word, "Cloud Computing," which is frequently used on TV and on the Internet. Do you know what the term means?

George: Well, to my knowledge, it simply means that the user can use storage, computing power, or specially crafted development environments without having to worry about how these work internally.

Simon: So what does "cloud" mean here in "Cloud Computing"?

George: Oh, that is a metaphor. It is a metaphor for the Internet based on how the Internet is described in computer network diagrams, which means it is an abstraction hiding the complex infrastructure of the Internet.

Simon: Can you further explain it to me? I still feel confused.

George: OK, let me explain it in this way. Cloud Computing is a style of computing in which IT-related capabilities are provided "as a service," allowing users to access technology-enabled services from the Internet ("in the cloud"). As customers generally do not own the infrastructure or know all the details about it, since they are mainly accessing or renting, they consume resources as a service, and may be paying for what they do not need, instead of what they actually do need to use.

Simon: Can you give me an example?

George: All right! Many cloud computing providers have adopted the utility computing model which is analogous to how traditional public utilities like electricity are consumed, which others are billed on a subscription basis. By sharing consumable and "intangible" computing power between multiple "tenants," utilization rates can be improved (as servers are not left idle) which can reduce costs significantly while increasing the speed of application development.

Simon: OK, I think I've got the points.

 Words & Expressions

Cloud Computing	云计算
craft [krɑːft]	v. 精巧地制作
metaphor ['metəfə]	n. 隐喻
diagram ['daɪəgræm]	n. 图表；示意图
abstraction [əb'strækʃ(ə)n]	n. 抽象化
infrastructure ['ɪnfrəstrʌktʃə]	n. 基础设施
analogous [ə'næləgəs]	adj. 相似的，模拟的
subscription [səb'skrɪpʃ(ə)n]	n. 认购；预订；订阅

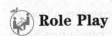 **Role Play**

A: Excuse me, Miss/Mrs/Mr... Have you come across any new items online recently?

B: Yes, I got to know...

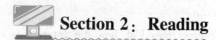 Section 2: Reading

---

Pre-reading Activities

1. Have you heard of Artificial Intelligence (AI)? Talk about the abilities of AI with your partner.

2. Do you know the goal in AI?

Artificial Intelligence

What is Artificial Intelligence (AI) exactly? As a beginning we offer the following definition:

AI is a branch of computer science concerned with the study and creation of computer systems that exhibit some form of intelligence: systems that learn new concepts and tasks, systems that can reason and draw useful conclusions about the world around us, systems that can understand a natural language or perceive and comprehend a visual scene, and systems that perform other types of feats that require human types of intelligence.

Like other definitions of complex topics, an understanding of AI requires an understanding of related terms such as intelligence, knowledge, reasoning, thought, cognition, learning and a number of computer-related terms. While we lack precise scientific definitions for many of these terms, we can give general definitions of them. And, of course, one of the objectives of this text is to impart special meaning to all of the terms related to AI, including their operational meanings.

Dictionaries define intelligence as the ability to acquire, understand and apply knowledge, or the ability to exercise thought and reason. Of course, intelligence is more than this. It embodies all of the knowledge and feats, both conscious and unconscious, which we have acquired through study and experience, highly refined sight and sound perception; thought; imagination; the ability to converse, read, write, drive a car, memorize and recall facts, express and feel emotions and much more.

Intelligence is the integrated sum of those feats which gives us the ability to remember a face not seen for thirty or more years, or to build and send rockets to the moon. It is those capabilities which set Homo sapiens apart from other forms of living things. And, as we shall see the food for this intelligence is knowledge.

Can we ever expect to build systems which exhibit these characteristics? The answer to this question is yes! Systems have already been developed to perform many types of intelligent tasks, and expectations are high for near term development of even more impressive systems.

We now have systems which can learn from examples, from being told, from past related experiences, and through reasoning. We have systems which can solve complex problems in mathematics, in scheduling many diverse tasks, in finding optimal system configurations, in planning complex strategies for the military and for business, in diagnosing medical diseases and other complex systems, to name a few. We have systems which can "understand" large parts of natural languages. We have systems which can see well enough to "recognize" objects from photographs, video cameras and other sensors. We have systems which can reason with incomplete and uncertain facts. Clearly, with these developments, much has been accomplished since the advent of the digital computer.

In spite of these impressive achievements, we still have not been able to produce coordinated, autonomous systems which possess some of the basic abilities of a three-year-old child. These include the ability to recognize and remember numerous diverse objects in a scene, to learn new sounds and associate them with objects and concepts, and to adapt readily to many diverse new situations. These are the challenges now facing researchers in AI. And they are not easy ones. They will require important breakthroughs before we can expect to equal the performance of our three-year old.

To gain a better understanding of AI, it is also useful to know what AI is not. AI is not the study and creation of conventional computer systems. Even though one can argue that all programs exhibit some degree of intelligence, an AI program will go beyond this in demonstrating a high level of intelligence to a degree that equals or exceeds the intelligence required of a human in performing some task.

AI is not the study of the mind, nor of the body, nor of languages, as customarily found in the fields of psychology, physiology, cognitive science, or linguistics. To be sure, there is some overlap between these fields and AI. All seek a better understanding of the human's intelligence and sensing processes.

But in AI the goal is to develop working computer systems that are truly capable of performing tasks that require high levels of intelligence. The programs are not necessarily meant to imitate human senses and thought processes. Indeed, in performing some tasks differently, they may actually exceed human abilities. The important point is that the systems are all capable of performing intelligent tasks effectively and efficiently.

Finally, a better understanding of AI is gained by looking at the component areas of study that make up the whole. These include such topics as robotics, memory organization, knowledge representation, storage and recall, learning models, inference techniques, commonsense reasoning, dealing with uncertainty in reasoning and decision making, understanding natural language, pattern recognition and machine vision methods, search and matching, speech recognition and synthesis, and a variety of AI tools.

How much success have we realized in AI to date? What are the next big challenges? The answers to these questions form a large part of the material covered in this text. We only mention here that AI is coming of an age where practical commercial products are now available including a variety of robotic devices, vision systems that recognize shapes and objects, expert systems that perform many difficult tasks as well as or better than their human expert counterparts, intelligent instruction systems that help pace a student's learning and monitor the student's progress, "intelligent" editors that assist users in building special knowledge bases, and systems which can learn to improve their performance.

📧 Words & Expressions

artificial [ˌɑːtɪˈfɪʃ(ə)l]	*adj.*	人造的；仿造的
intelligence [ɪnˈtelɪdʒ(ə)ns]	*n.*	智力
definition [defɪˈnɪʃ(ə)n]	*n.*	定义
concept [ˈkɒnsept]	*n.*	观念，概念
perceive [pəˈsiːv]	*vt.*	察觉，感觉；理解；认知
feat [fiːt]	*n.*	技艺表演
reasoning [ˈriːz(ə)nɪŋ]	*n.*	推理；论证；评理 *adj.* 推理的
cognition [kɒgˈnɪʃ(ə)n]	*n.*	认识；知识；认识能力
precise [prɪˈsaɪs]	*adj.*	精确的；明确的；严格的
impart [ɪmˈpɑːt]	*vt.*	给予（尤指抽象事物），传授；告知，透露
embody [ɪmˈbɒdɪ]	*vt.*	体现，使具体化；具体表达
refine [rɪˈfaɪn]	*vt.*	精炼，提纯；改善；使……文雅
perception [pəˈsepʃ(ə)n]	*n.*	知觉；[生理] 感觉；看法；洞察力；获取
converse [kənˈvɜːs]	*adj.*	相反的，逆向的，颠倒的
integrated [ˈɪntɪgreɪtɪd]	*adj.*	综合的；完整的；互相协调的
Homo sapiens		智人（现代人的学名）；人类
optimal [ˈɒptɪm(ə)l]	*adj.*	最佳的；最理想的
military [ˈmɪlɪt(ə)rɪ]	*adj.*	军事的；军人的；适于战争的
diagnose [ˈdaɪəgnəʊz]	*vt.*	诊断；断定
sensor [ˈsensə]	*n.*	传感器
accomplish [əˈkʌmplɪʃ]	*vt.*	完成；实现；达到
advent [ˈædvənt]	*n.*	到来；出现
coordinated [kəʊˈɔːdɪneɪtɪd]	*vt.*	调整；使调和（coordinate 的过去分词）；调节；整理 *adj.* 协调的
autonomous [ɔːˈtɒnəməs]	*adj.*	自治的；自主的；自发的
breakthrough [ˈbreɪkθruː]	*n.*	突破；突破性进展
demonstrate [ˈdemənstreɪt]	*vt.*	证明；展示；论证
customarily [ˈkʌstəm(ə)rɪlɪ]	*adv.*	通常，习惯上
psychology [saɪˈkɒlədʒɪ]	*n.*	心理学
physiology [ˌfɪzɪˈɒlədʒɪ]	*n.*	生理学
linguistics [lɪŋˈgwɪstɪks]	*n.*	语言学
overlap [əʊvəˈlæp]	*n.*	重叠；重复
inference [ˈɪnf(ə)r(ə)ns]	*n.*	推理；推论；推断
synthesis [ˈsɪnθɪsɪs]	*n.*	综合，[化学] 合成；综合体

Unit 9　New Technologies

🤖 Basic Technical Terms

Artificial Intelligence	人工智能
sensor	传感器
robotic	机器人
memory organization	存储组织
inference technique	推理技术
commonsense reasoning	常识推理

Choose the best answer according to the text.

(　) 1. AI is _____.
　　A. a branch of computer science　　B. a branch of psychology
　　C. a branch of physiology　　D. a branch of linguistics

(　) 2. As we shall see the food for this intelligence is _____.
　　A. data　　B. electricity　　C. knowledge　　D. vitamin

(　) 3. Before we can expect to equal the performance of our three-year-old, which of the following is not a challenge?
　　A. The ability to recognize and remember.
　　B. The ability to learn new sounds.
　　C. The ability to adopt readily.
　　D. The ability to perform some simple intelligent tasks.

(　) 4. Which of the following is an easy challenge for AI?
　　A. To recognize and remember numerous diverse objects in a scene.
　　B. To learn new sounds and associate them with objects and concepts.
　　C. To adapt readily to many diverse new situations.
　　D. To see well enough to "recognize" objects from photographs, video cameras and other sensors.

(　) 5. Which of the following sentences is NOT right?
　　A. AI is a branch of computer science concerned with the study and creation of computer systems that exhibit some form of intelligence.
　　B. Like other definitions of complex topics, an understanding of AI requires an understanding of related terms such as Central Processing Unit.
　　C. To gain a better understanding of AI, it is also useful to know what AI is not.
　　D. A better understanding of AI is gained by looking at the component areas of study that make up the whole.

Section 3：Computer Terms

Do You Know?

Artificial Intelligence：人工智能，英文缩写为 AI。它是研究、开发用于模拟、延伸和扩展人的智能的理论、方法、技术及应用系统的一门新的技术科学。人工智能是计算机科学的一个分支，它企图了解智能的实质，并生产出一种新的能以人类智能相似的方式做出反应的智能机器，该领域的研究包括机器人、语言识别、图像识别、自然语言处理和专家系统等。人工智能从诞生以来，理论和技术日益成熟，应用领域也不断扩大，可以设想，未来人工智能带来的科技产品，将会是人类智慧的"容器"。人工智能可以对人的意识、思维的信息过程进行模拟。人工智能不是人的智能，但能像人那样思考，也可能超过人的智能。

Virtual Reality：虚拟现实技术，是仿真技术的一个重要方向，是仿真技术与计算机图形学、人机接口技术、多媒体技术、传感技术、网络技术等多种技术的集合，是一门富有挑战性的交叉技术前沿学科和研究领域。虚拟现实技术（VR）主要包括模拟环境、感知、自然技能和传感设备等方面。模拟环境是由计算机生成的、实时动态的三维立体逼真图像。感知是指理想的 VR 应该具有一切人所具有的感知。除计算机图形技术所生成的视觉感知外，还有听觉、触觉、力觉、运动等感知，甚至还包括嗅觉和味觉等，也称为多感知。自然技能是指人的头部转动，眼睛、手势或其他人体行为动作，由计算机来处理与参与者的动作相适应的数据，并对用户的输入作出实时响应，并分别反馈到用户的五官。传感设备是指三维交互设备。

Smart Home：智能家居，是利用先进的计算机技术、网络通信技术、智能云端控制、综合布线技术、医疗电子技术依照人体工程学原理，融合个性需求，将与家居生活有关的各个子系统如安防、灯光控制、窗帘控制、煤气阀控制、信息家电、场景联动、地板采暖、健康保健、卫生防疫、安防保安等有机地结合在一起，通过网络化综合智能控制和管理，实现"以人为本"的全新家居生活体验。

Cloud Computing：云计算，是基于互联网的相关服务的增加、使用和交互模式，通常涉及通过互联网来提供动态易扩展且经常是虚拟化的资源。云是网络、互联网的一种比喻说法。过去在图中往往用云来表示电信网，后来也用来表示互联网和底层基础设施的抽象。因此，云计算甚至可以让你体验每秒 10 万亿次的运算能力，拥有这么强大的计算能力可以模拟核爆炸、预测气候变化和市场发展趋势。用户通过计算机、手机等方式接入数据中心，按自己的需求进行运算。

Quantum Computation：量子计算，是一种遵循量子力学规律调控量子信息单元进行计算的新型计算模式。对照于传统的通用计算机，其理论模型是通用图灵机；通用的量子计算机，其理论模型是用量子力学规律重新诠释的通用图灵机。从可计算的问题来看，量子计算机只能解决传统计算机所能解决的问题，但是从计算的效率上，由于量子力学叠加性的存在，目前某些已知的量子算法在处理问题时速度要快于传统的通用计算机。

Block Chain：区块链，是用分布式数据库识别、传播和记载信息的智能化对等网络，也称为价值互联网。中本聪在 2008 年，于《比特币白皮书》中提出"区块链"概念，并在 2009 年创立了比特币社会网络，开发出第一个区块，即"创世区块"。

IOT：物联网（Internet of Things 的缩写），是互联网、传统电信网等信息承载体，让所有能行使独立功能的普通物体实现互联互通的网络。物联网一般为无线网，由于每个人周围的设备可以达到 1 000～5 000 个，所以物联网可能要包含 500～1 000 兆个物体。在物联网上，每个人都可以应用电子标签将真实的物体上网联结，在物联网上都可以查出它们的具体位置。通过物联网可以用中心计算机对机器、设备、人员进行集中管理、控制，也可以对家庭设备、汽车进行遥控，以及搜索位置、防止物品被盗等，类似自动化操控系统，同时通过收集这些小事的数据，最后可以聚集成大数据，包含重新设计道路以减少车祸、都市更新、灾害预测与犯罪防治、流行病控制等社会重大改变，实现物和物相联。物联网的应用领域主要包括以下方面：运输和物流领域、工业制造领域、健康医疗领域、智能环境（家庭、办公、工厂）领域、个人和社会领域等，具有十分广阔的市场和应用前景。

Section 4：Exercises

Ⅰ. Tell whether the following statements are True（T）or False（F）.

(　　) 1. AI is a branch of computer science concerned with the study and creation of computer systems that exhibit some form of intelligence.

(　　) 2. While we lack precise scientific definitions for many of these terms, we can still give exact definitions of them.

(　　) 3. Scientists define intelligence as the ability to acquire, understand and apply knowledge, or the ability to exercise thought and reason.

(　　) 4. It is Intelligence which sets Homo sapiens apart from other forms of living things.

(　　) 5. In spite of these impressive achievements, we are able to produce coordinated, autonomous systems which possess some of the basic abilities of a three-year-old child.

(　　) 6. AI is not the study and creation of conventional computer systems.

(　　) 7. AI is the study and creation of conventional computer systems.

(　　) 8. The AI programs are necessarily meant to imitate human senses and thought processes.

(　　) 9. A better understanding of AI is gained by looking at the component areas of study that make up the whole.

(　　) 10. AI is coming of an age where practical commercial products are now available.

Ⅱ. Complete the following sentences according to the text.

1. AI is a branch of computer science concerned with the study and creation of computer systems that exhibit some form of intelligence：systems that _____, systems that _____, systems that _____, and systems that _____.

2. An understanding of AI requires an understanding of related terms such as _____,
_____, _____, _____, _____, _____ and a number of _____.
3. Dictionaries define intelligence as the ability to _____, _____ and _____ knowledge, or the ability to exercise _____ and _____.
4. AI is not the study of the _____, nor of the _____, nor of _____, as customarily found in the fields of _____, _____, _____, or _____.
5. The programs are not necessarily meant to imitate human _____ and _____ processes.

Ⅲ. **Ability to explore.**
1. Which of the following is NOT AI software?
   A. DEEP BLUE   B. Alpha Go   C. Siri   D. Windows
2. Do you know the differences between BOTTOM-UP AI（强人工智能）and TOP-DOWN AI（弱人工智能）?

Ⅳ. **Fill in the table below by giving the corresponding translation with the help of dictionaries or Internet.**

English	Chinese
Machine Learning	
	自然语言处理
Affective Computing	
	机器人学
Machine Perception	
	机器认知
Knowledge Representation	

Section 5：Furthering Reading

RealCine-virtual Reality for Everyone

This presentation will give you some information about RealCine：how it works, why it is better than a film, and how it can be used in other ways. The RealCine experience will amaze you, and you will agree that this is an extraordinary technology that deserves to be developed further.

The technology behind RealCine is virtual reality (VR). Unlike a film, where a passive audience watches and hears what is happening on a screen, RealCine puts you into the action and connects with your senses of sight, hearing, smell and touch in an active way. Imagine that a VR user goes sightseeing in the Himalayas. Not will he or she feel every step of climbing Mount

Unit 9 New Technologies

Qomolangma, but the user will also experience the cold, smells, sights and sounds of the surrounding environment; he or she will enjoy a feeling of happiness and a sense of achievement upon reaching the top.

RealCine works by making the users feel that they are really in a new world—a world that does not exist except in a computer program. To achieve this, special VR headsets are designed to allow the users to see in 3D and hear the sound all around them. The movements of the headset indicate the direction in which the user wants to go. The user also wears special gloves so he or she can "touch" the people and objects that he or she sees. To add to the virtual world of RealCine, the headsets even have small openings that give out smells to match the environment. Both the headsets and the gloves are connected to a computer network in the VR studio.

In scientific studies it has been shown that VR can be a good treatment for people who have social problems. In one case, a teenager who was afraid of talking and playing with his schoolmate was treated with VR. In the world created by RealCine, he became the captain of the Brazilian football team and scored the winning goal in a World Cup final. This encouraged him to become more confident around others.

An argument has been put forward that some users will be disappointed by RealCine because VR is not real. However, with VR we are able to do things that could never be achieved in real life. For example, with the aid of RealCine, a seventy-year-old grandfather recently took a trip to Africa. In reality, he is disabled and can no longer walk, but he was able to see and touch a lion while still in the convenience of the VR studio.

Besides this, VR can be used to practice skills in a secure environment that otherwise would be quite dangerous. For example, firefighters could use RealCine to train safely, without the risk of getting injured in a burning building. It could be used in class as well. Teachers could bring history alive by placing students in an ancient town, or they could teach biology by allowing students to experience the world as a whale or a squirrel.

Finally, RealCine provides fantastic technology for urban planning. Engineers can enter the design of a neighborhood into a computer, and then use VR to "walk" around the neighbourhood, see how it looks and make changes before construction is carried out. This kind of urban planning is in the long term cheaper and more practical, compared with the way most urban planning is done today. I recommend the government use this technology in the future planning of this city.

How well do you understand the business presentation? Read it again and answer the following questions.

Answer the following questions according to the passage.

1. What is the name of the product?

2. What are the advantages of using RealCine for urban planning?

3. What technology is behind this product?

4. What do users wear so they feel that they are really in a new world?

5. How could firefighters be trained with this new technology?

Self-checklist

根据实际情况，从 A、B、C、D 中选择合适的答案：A 代表你能很好地完成该任务；B 代表你基本上可以完成该任务；C 代表你完成该任务有困难；D 代表你不能完成该任务。

A　　B　　C　　D

☐　☐　☐　☐　1. 理解并正确朗读听说部分的句子，正确掌握发音和语调。

☐　☐　☐　☐　2. 模仿对话部分的句型进行简单的对话。

☐　☐　☐　☐　3. 读懂本课的短文，并正确回答相关问题。

☐　☐　☐　☐　4. 向同学介绍信息新技术的相关内容。

☐　☐　☐　☐　5. 掌握并能运用本单元所学重点句型、词汇和短语。

☐　☐　☐　☐　6. 使用电子词典查阅 Furthering Reading，进一步了解人工智能。

Unit 10

Job Application

> **Unit Goals**
>
> In this unit, we are focusing on the following issues:
> ★ Identify steps for job application documents
> ★ Be able to write a job application and a resume for yourself
> ★ Be able to file relative materials before seeking jobs

 Warm-up

Match the pictures with the words and phrases, then write the correct letter in the bracket to each.

(   ) 1. Resume   (   ) 2. Letter of application

(   ) 3. Job advertisement   (   ) 4. Interview

A

B

C   D

— 85 —

Section 1: Dialogue

Directions: Read the following dialogue in pairs and talk about your career plan.
Situation: *The dialogue is between an IT company employer, George, and a new employee, Simon, about Simon's working experience.*

Working in an IT Company

Simon: Do you have any experience of working in IT companies?
George: Yes, I used to have a part-time job in an IT company.
Simon: What was your main duty there?
George: Computer programming.
Simon: What qualities do you think a computer programmer should have?
George: In my opinion, a computer programmer should have teamwork spirit and active imagination. He or she should be talented in creation and sensitive to the changes in the IT Market.
Simon: What have you learned from the part-time job you have had?
George: I have learned how to behave myself as a qualified employee and some skills about how to get along with people with different personalities.
Simon: Could you give me a detailed explanation?
George: Yes, I've learned from my working experience that sometimes the working situation may become pretty tense. For example, people may get impatient because deadlines are always there. This sometimes causes people to get upset or say something they usually wouldn't say. When this happens, I just let it go and keep working because I know things will eventually calm down.

✉ Words & Expressions

part-time job	兼职工作
main [meɪn] **duty** [ˈdjuːtɪ]	主要工作职责
computer programming	计算机编程
teamwork spirit	团队合作精神
deadline [ˈdedlaɪn]	*n.* 最后期限

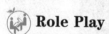 **Role Play**

A: Excuse me, Miss/Mrs/Mr... What kind of job do you want to do in the future?
B: Well, I should say...

Unit 10 Job Application

 Section 2: Reading

Pre-reading Activities

1. Have you heard of the following jobs? Talk with your partner about its responsibilities.
   programmer   computer operator   computer engineer
   hardware engineer   software engineer   network maintenance engineer
2. How to write a job application letter?

The Letter of Application Sample

102 River Road
Nanjing 210000
(025) 22244556
December 30, 2018

Mr. Mike Anderson
Director of Human Resources
ABC Company
168 Center Street
Nanjing 210000

Dear Mr. Anderson,

  I am writing in reply to the enclosed advertisement in Yangtse Evening on December 27 for a computer operator. I beg to offer as a candidate for the post.

  I am 20 years old and will graduate from Jiangsu Union Technical Institute. My major in college is just computer Science and Technology. The major courses I have taken include Computer Program Designing, Auto-CAD, Visual Basic, Network and Communications, OA Equipment, and so on. During the four-year studies, I have always belonged to the top in class academically, and have won the title of "Excellent Student Leader" twice and first-class scholarship three times. Besides, I got the certificates of PETS - 3 and Computer Stage 2. What's worth mentioning is that I joined the Computer Image & Word Processing Studio in our school when I was a sophomore. What I did in the studio is to edit pictures and files. After a year in the studio, I joined the Network Center in our school. Although I was just a coder there, but I learned a lot and I knew the process of how to program a website with C# and ASP. NET. According to the four years' learning and practice, I think I have the competence that you need.

> I have an inquisitive and analytical mind. I enjoy finding out about things. I have tact and good humour and the ability to draw people out.
>
> If these meet your requirements, please grant me an interview.
>
> I have enclosed my personal resume, a recent photo and a couple of copies of my certificates as requested.
>
> I look forward to hearing from you.
>
> Yours sincerely,
Wang Bing
December 30, 2018

Words & Expressions

reply [rɪˈplaɪ]	v. 答复；回应 n. 回答；答复
enclosed [ɪnˈkləʊzd]	adj. 被附上的
advertisement [ədˈvɜːtɪsmənt]	n. 广告，宣传；公告；出公告，做广告
operator [ˈɒpəreɪtə]	n. 经营者；操作员
candidate [ˈkændɪˌdeɪt]	n. 候选人；应试者
institute [ˈɪnstɪtjuːt]	n. 协会；学会；学院
communication [kəˌmjuːnɪˈkeɪʃən]	n. 交流；传达；通信；沟通
equipment [ɪˈkwɪpm(ə)nt]	n. 设备，装备；器材
academically [ˌækəˈdemɪkəlɪ]	adv. 从学术观点看
certificate [səˈtɪfɪkɪt]	n. 证书；执照，文凭
sophomore [ˈsɒfəmɔː]	n. 大学（或中学）二年级学生
inquisitive [ɪnˈkwɪzɪtɪv]	adj. 好奇的；爱打听的；求知欲强的
analytical [ˌænəˈlɪtɪkəl]	adj. 分析的，分析法的；善于分析的
tact [tækt]	n. 机智，机敏；老练
grant [grɑːnt]	vt. 授予；允许；承认 vi. 同意
resume [rɪˈzjuːm]	n. 摘要，[管理] 履历，简历

Basic Technical Terms

Computer Program Designing	计算机程序设计
AutoCAD	计算机绘图
Visual Basic, VB	一种通用的基于对象的程序设计语言
Network and Communications	网络与通信
OA Equipment	办公自动化设备

Read the passage and answer the following questions.

1. What position did Wang Bing apply for?

2. Where did Wang Bing see the post?
   _____
3. Which school did he graduate from?
   _____
4. Did he have some relevant working experience?
   _____

Section 3: Computer Terms

Do You Know?

常用表达：与计算机行业有关的职位名称
Programmer 程序员
Computer Operator 计算机操作员
Hardware Engineer 硬件工程师
Project Manager 项目经理
Technical Engineer 技术工程师
Systems Engineer 系统工程师
Computer Technician 计算机技术员
Developmental Engineer 开发工程师
LAN Administrator 局域网管理员
Systems Analyst 系统分析员
Product Support Manager 产品支持经理

常用表达：个人情况

nationality 民族	native place 籍贯	health 健康状况
height 身高	ID number 身份证号	marital status 婚姻状况
weight 体重	birth date 出生日期	single 未婚的
blood type 血型	married 已婚	sex 性别
male 男	female 女性（的）	

常用表达：教育

major 主修	minor 辅修	vocation job 假期工作
part-time job 兼职	scholarship 奖学金	student union 学生会
social practice 社会实践	social activities 社会活动	excellent leader 优秀干部
in-job training 在职培训	academic year 学年	supervisor 论文导师
mark 分数	degree 学位	bachelor 学士
master 硕士	doctor 博士	post doctorate 博士后

abroad student 留学生   undergraduate 在校生   graduate 毕业生

常用表达：工作经历

design and implement 设计与实现   train and supervise 培训与指导
responsible for 负责   assist in 辅助   test 测试
develop 开发   accomplish 完成   direct 指导
design 设计   be promoted to 被提升为

常用表达：特长

versed in 精通   special expertise in 专长于   excellent…skill 擅长……技能
proficient in 精通   hobby 业余爱好   long distance running 长跑
collecting stamps 集邮   traveling 旅游   listening to symphony 听交响乐

Section 4：Exercises

Ⅰ. Complete the following form with the information you get from the passage.

Name：_____   Address：_____

Item	Information in an application letter
Education background	1. Graduate from：_____ 2. Standard tests：_____
Relevant courses	
Work experience	Part-time jobs in school： 1. _____ 2. _____
Relevant skills	
Awards and achievements	

Ⅱ. Put the verbs in the passage into correct tenses.

After university you will start _____ (hunt) for a job, because _____ (seek) a job can be very difficult nowadays. If you want to _____ (apply) for a job advertising in a newspaper or magazine, first you must write an _____ (apply) letter stating that you are interested in it. You then need to fill in an _____ (apply) form and attach your CV. The CV, similar to a _____ (resume) in American English, is the paper stating your working experience or education. If the company needs a reference, you have to ask your boss or professor to be a _____ (refer).

III. Ability to explore.

Read the following ten sentences and finish part 1 and part 2.

A. Besides accounting, I have also studied Maths, English, tax law and some other subjects.
B. I'm a student from a Shanghai vocational school and I will graduate next summer.
C. Though I'm young and lack full-time working experience, I'm hard-working and responsible.
D. Please contact me at 13424357899 or e-mail cathy678@126.com.
E. Your prompt reply would be greatly appreciated.
F. I'm writing to apply for the position of Assistant Accountant in the Financial Department of your company.
G. I sincerely hope that you could give me an interview.
H. I study very hard and always get high marks for each subject.
I. I worked as a part-time assistant accountant in my relative's company.
J. My major is accounting.

Part 1

1. Tell which sentences show the purpose of writing the letter: _____

2. Tell which sentences show the education: _____

3. Tell which sentences show the work experience: _____

4. Tell which sentences show the personality: _____

5. Tell which sentences show the contact information: _____

6. Tell which sentences express the hope: _____

Part 2

Put the above sentences in a logical order in order to form the body of an application letter.

_____

IV. In order to apply for a job, you need to fill out the CV. Fill in the form by yourself.

Name（姓名）	English（英文）		Attach a photo taken within 12 months 请贴最近一年内所拍的照片
	Chinese（中文）		
Birth Date（出生日期）		Height（身高）	
Birth Place（出生地点）		Weight（体重）	
Nationality（民族）		Male（男）	
ID Card No.（身份证号码）		Female（女）	

续表

Address（通信地址）			Tel（电话）	
Education Background（教育程度）				
Grade（等级）	Name of School（学校名称）	From（自）	To（至）	
Primary（小学）		Mo.（月） Yr.（年）	Mo.（月） Yr.（年）	
Secondary（初中）				
Institute（中专或大专）				
Others（其他）				
Technical Qualifications（技术资格）				
Extracurricular Activities（课外活动）				
Part-time Jobs（业余工作）				
Special Skills（特别技能）				
Interests and Hobbies（兴趣和爱好）				
Rewards（奖励）				

Ⅴ. **Fill in the table below by giving the corresponding translation with the help of dictionaries or Internet**.

English	Chinese
Application letter	
	面试
CV	
	推荐人
Courses taken	
	必修课程
Specialized courses	
	三好学生
major in	
	学士学位

Section 5：Furthering Reading

Pre-reading Activity

Do you know the meaning of "resume"? Before job hunting, is it important to design your resume, and how to write it?

Unit 10 Job Application

Resume

Zhang Yuxuan
28 Zhongshan Road
Nanjing, Jiangsu Province 210000
P. R. China
Telephone: 025 - 86775834
E-mail: yuxuan@163.com
Position applied for
System Engineer

Education
2012 - 2016 Computer Science at Nanjing University of Technology
2009 - 2012 Tianjiabing High school in Changzhou, Jiangsu Province, P. R. China

Oct. 2016 - present
Employee at Beyond Co., Ltd.
Position: Senior System Engineer
Working Environment
    Operating System: PC's with Unix and Linux
    Programming Language: C
    Script Language: Korn Shell, etc.
Projects
    Modifying Customer Care Server (CCS)
    Developing Customer Care Client Emulator (CCCE)
    Developing Test-Script of Customer Care Server (test_ccs)
    Writing Documentation of test_ccs

Oct. 2012 - July. 2016
Employed by DUT Computer Center
Position: Student Assistance in Computer Room
Working Environment
    Programming Language: Java/J++, C/C++, HTML, X-Window
    Operating System: Unix, Linux, Windows NT/9x/3.1, Mac OS

 Answer the following questions according to the passage.

1. What is the position applied for?

2. Which school did Zhang Yuxuan graduate from? What is his major?

3. What kind of working experience does he have?

4. How many languages can he program with? What are they?

Self-checklist

根据实际情况，从 A、B、C、D 中选择合适的答案：A 代表你能很好地完成该任务；B 代表你基本上可以完成该任务；C 代表你完成该任务有困难；D 代表你不能完成该任务。

A	B	C	D	
☐	☐	☐	☐	1. 理解并正确朗读听说部分的句子，正确掌握发音和语调。
☐	☐	☐	☐	2. 模仿听说部分的句型进行简单的对话。
☐	☐	☐	☐	3. 读懂本课的短文，并正确回答相关问题。
☐	☐	☐	☐	4. 仿照课文，制作个人求职信和简历。
☐	☐	☐	☐	5. 掌握并能运用本单元所学重点句型、词汇和短语。
☐	☐	☐	☐	6. 使用电子词典查阅 Furthering Reading，了解求职的相关信息。

参考译文

第二单元 计算机硬件组成

当我们说起计算机时,这种形象(图1)就会显现在我们的脑海中:作为基本输出装置的显示屏、作为基本输入装置的一个键盘和一个鼠标、被称为机箱的一个机器盒子。

计算机硬件可以被分为4个部分:中央处理器、存储装置、输入装置和输出装置。

中央处理器

中央处理器(CPU)是计算机的大脑。CPU 的设计,除了影响计算机能够有效使用的主要存储量,也影响着计算机的处理能力和速度。

存储装置

我们通常把存储装置分为2种类型:主存储器和二级或辅助存储器。存储器被分为 RAM(随机存取存储器)和 ROM(只读内存)。大多数计算机的主存储器由 RAM 组成。我们可以把资料和程序储存到 RAM。你计算机中 RAM 的大小直接影响到你能使用的软件的复杂程度。当计算机关机的时候,主存储器是空的。ROM 可以被读取,但不能被写入。当计算机关机的时候,存储在 ROM 中的指令不会丢失。硬盘驱动和 U 盘是二级存储器的常见种类。

输入装置

最常见的输入装置就是键盘和鼠标。键盘上的键让你在计算机中输入信息和指令。鼠标可以让你选择并移动屏幕上的项目。

输出装置

最常见的输出装置是显示器和打印机。显示器显示文本和图像。打印机生成显示在屏幕上的信息的纸质副本。喷墨打印机和激光打印机是当前计算机市场上常见的两种打印机。

第三单元 操作系统

众所周知,计算机系统大致可以分为四个部分:硬件、操作系统、应用软件和用户

（图1）。

硬件提供了基本的计算资源。应用程序定义了这些资源用于解决用户计算问题的方式。可能有许多不同的用户试图解决不同的问题，所以可能有很多不同的应用程序。

操作系统（OS）是管理计算机硬件和软件资源并为计算机程序提供公共服务的系统软件。它充当计算机用户和计算机硬件之间的接口。操作系统提供了一个用户可以在其中执行程序的环境。因此，操作系统的主要目标是使计算机系统易于使用。第二个目标是以一种有效的方式使用计算机硬件。我们可以将操作系统视为资源分配器。计算机系统有许多解决问题可能需要的资源：CPU时序、内存空间、文件存储、输入/输出（I/O）设备，等等。

许多不同种类的操作系统可以安装在各种设备上。根据应用领域，操作系统可以分为三个主要类别：桌面操作系统、服务器操作系统和嵌入式操作系统。

桌面操作系统

桌面操作系统主要用于个人计算机上。个人计算机市场从硬件架构上来说主要分为两大阵营，即PC机与Mac机；从软件上来说主要分为两大类，即类Unix操作系统和Windows操作系统：

1. Unix和类Unix操作系统：Mac OS X，Linux发行版（如Debian，Ubuntu，Linux Mint，openSUSE，Fedora等）；

2. 微软公司Windows操作系统：Windows XP，Windows Vista，Windows 7，Windows 8等。

服务器操作系统

服务器操作系统一般指的是安装在大型计算机上的操作系统，比如Web服务器、应用服务器和数据库服务器等。服务器操作系统主要集中在三大类：

1. Unix系列：SUNSolaris，IBM－AIX，HP－UX，FreeBSD等；

2. Linux系列：Red Hat Linux，CentOS，Debian，Ubuntu等；

3. Windows系列：Windows Server 2003，Windows Server 2008，Windows Server 2008 R2等。

嵌入式操作系统

嵌入式操作系统是应用在嵌入式系统的操作系统。

嵌入式系统广泛应用在生活的各个方面，涵盖范围从便携设备到大型固定设施，如数码相机、手机、平板计算机、家用电器、医疗设备、交通灯、航空电子设备和工厂控制设备等。

在嵌入式领域常用的操作系统有嵌入式Linux、Windows Embedded、VxWorks等，以及广泛使用在智能手机或平板计算机等电子产品的操作系统，如Android、iOS、Symbian、Windows Phone和BlackBerry OS等。

第四单元　应用软件

软件是程序的另一个名称。众所周知，没有软件的支持，计算机什么也做不了。如果我们把计算机描述为一个人，我们通常说硬件就像一个人的身体，而软件就像灵魂。就像一个人有很多想法一样，计算机有好几种软件，每种都用来做不同的工作。软件可以分为两大类：系统软件和应用软件。没有前者，计算机就无法运行。如果没有后者，计算机——不管它有多强大——都不会对你的业务有多大帮助。

有了系统软件，当你打开计算机的时候，你就可以让它运转起来：把信息写到磁盘上，检查病毒和做其他一些活动。应用软件可由客户制作或包装。它做最终用户工作。

应用软件有多种类型：

应用程序套件由多个捆绑在一起的应用程序组成。它们通常具有相关的功能、特性和用户界面，并且可以相互交流，例如打开彼此的文件。

业务应用程序通常以套件的形式出现，例如 Microsoft Office、OpenOffice.org 和 iWork，它们将文字处理器、电子表格等捆绑在一起；但套件的存在是为了其他目的，例如图形或音乐。

企业软件解决了组织过程和数据流的需求，通常是在大型分布式环境中运作。

信息工作者软件满足个人创建和管理信息的需要。

教育软件与内容访问软件相关，但其内容和/或功能适合教育者或学生使用。

仿真软件是为研究、训练或娱乐目的而对物理或抽象系统进行仿真的计算机软件。

媒体开发软件可以满足为他人提供打印和电子媒体的个人的需求，通常在商业或教育环境中使用。

移动应用程序运行在手持设备上，如移动电话、个人数字助理和企业数字助理。

产品工程软件用于开发硬件和软件产品。这包括计算机辅助设计（CAD）、计算机辅助工程（CAE）、计算机语言编辑和编译工具、集成开发环境和应用程序编程接口。

第五单元　多媒体

多媒体是使用由不同内容形式组合起来的媒体和内容。它不同于使用传统形式的印刷或手工制作材料的介质。多媒体包括文本、音频、静止图像、动画、视频和交互内容形式的组合。

在这四十年间，这个词的词义经历了不同阶段的演变。在 20 世纪 70 年代后期，这个术语被用来描述多个投影仪在音频的节奏下进行幻灯片放映而形成的演示过程。然而，到了 20 世纪 90 年代，"多媒体"又呈现出其全新的意义。

多媒体可以大致分为线性和非线性两类。线性主动的内容是在没有给观看者任何导航控制的情况下进行的，例如电影的放映。非线性内容则提供用户交互性，以控制与计算机的进度或用于基于自定进度的计算机学习训练。多媒体演示可以是实时的或录制的。录制的演示也可以允许通过导航系统进行交互。实时多媒体演示可以通过与演示者或表演者的交互来实现交互。

多媒体演示可以在舞台上亲自观看、投影、传输或在媒体播放器本地播放。广播电视可以是实时的或录制的多媒体演示。广播和录音采用的可以是模拟或数字电子媒体技术。人们可以下载或以流媒体传输数字在线多媒体资源。流媒体传输的多媒体也可以采用直播或

点播。

各种格式的技术或数字多媒体应用旨在增强用户的体验，例如使其更容易和更快地传达信息，或者在娱乐或艺术领域中，使其超越日常生活中的体验。

通过组合多种形式的媒体内容，可以实现增强的交互水平。在线多媒体越来越多地成为受众导向和数据驱动的应用程序，使应用程序能够实现和终端用户在多种内容上的创新合作和个性化。除了视听之外，触感技术还可以使用户感受到虚拟物体。涉及味觉和嗅觉幻觉的新兴技术也可以增强多媒体体验。

第六单元　什么是互联网及它是如何工作的？

因特网是一个互相连接的计算机网络的全球系统，它用标准的互联网协议组（TCP/IP）服务于全世界几十亿人。它是一个由数百万个从本地到全球范围的私人、公共、学术、商业和政府网络组成的网络，通过一系列广泛的电子和光学网络技术进行连接。互联网承载着广泛的信息资源和服务，如万维网（WWW）内相互链接的超文本文档和支持电子邮件的基础设施。

如果你想上网，你需要做什么？

如果你知道地址，只需小心地将其输入屏幕上的框中。如果你不知道地址，去找一个流行的搜索引擎，如谷歌（www.Google.com）或百度（www.baidu.com）。一旦到了那里，你只需要输入一些关键字。例如，你可能想了解你所在城市的天气。如果你住在南京，你会输入"南京"和"天气"。然后你会得到一个站点列表。选择正确的网站，你就可以得到你想要的信息。

然而，建议你不要相信网上的一切。当你使用互联网上的信息时，你必须小心。任何人都可以创建一个网站，所以你不能总是确保信息是正确的。由大公司或组织制作的官方网站可能比由非专家创建的网站更有用。通常最好使用多个信息源来确保信息是正确的。

第七单元　计算机安全

计算机安全是计算机与网络领域的信息安全的一个分支。其目的是在保证信息和财产可由被授权用户正常获取和使用的情况下，保护此信息和财产不受偷窃、污染、自然灾害等的损坏。

计算机系统安全是指一系列包含敏感和有价值的信息和服务的进程和机制，不被未得到授权和不被信任的个人、团队或事件公开、修改或损坏。由于它的目的在于防止不需要的行为发生而非使得某些行为发生，其策略和方法常常与其他大多数的计算机技术不同。

计算机安全技术的基础是逻辑学。安全性并非大部分计算机应用的主要目的，而在设计时就考虑程序的安全性常常会对程序的运行有所限制。

计算机病毒是一组可以复制和感染计算机的计算机程序。"计算机病毒"一词有时被作为一个包罗万象的术语用以包含所有类型的恶意软件，甚至是那些不具有繁殖能力的恶意软件。恶意软件包含计算机病毒、计算机蠕虫、特洛伊木马程序、大多数的Rootkit、间谍软件、虚假广告信息和其他的恶意和不需要的软件，以及真正的病毒。

正如人体的病毒侵害活细胞，然后将它变成类似制造病毒的工厂一样，计算机病毒是一个小程序，通过将该程序的本身附加到另一个程序上来进行复制。病毒一旦附加在主程序上，就会将另一些程序锁住令其感染。这样，如果病毒传染到了局域网或多用户系统的话，那么该病毒就会迅速传播到整个硬盘或整个（计算机的）组成部分。

为了保证计算机的安全，我们必须积极有意识地采取一些措施。
（1）在将未知来源的磁盘插入计算机时要非常小心。
（2）在操作任何文件之前，都要扫描磁盘的所有文件。
（3）仅从信誉良好的网站下载 Internet 文件。
（4）不要打开陌生人的电子邮件附件（尤其是可执行文件）。

现实应用中有这么多的反病毒软件和其他防护措施。许多用户在计算机上下载并安装了可以检测和消除已知病毒的反病毒软件。反病毒软件不改变主机软件传播病毒的潜在能力。用户必须定期更新他们的软件来修补安全漏洞。反病毒软件也需要定期更新，以识别最新的威胁。

第八单元　电子商务

电子商务，又称 e-business，是"electronic commerce"的缩写。

那么，什么是 electronic commerce 呢？

电子商务的含义在过去的 30 年里一直在变化。起先，电子商务意味着利用电子资金转移（EFT）等技术，使商业交易简易化。后来，信用卡、自动柜员机（ATM）和电话银行的发展也是电子商务的一种形式。现在我们说电子商务是指通过在线服务或互联网购买或销售产品的活动。它采用了多种技术，如移动商务、电子资金转移、供应链管理、互联网营销、在线交易处理、电子数据交换（EDI）、库存管理系统和自动数据收集系统。

一些研究人员根据参与交易或业务流程的实体的类型对电子商务进行分类。常见的五大类是企业对消费者、企业对企业、消费者对消费者、企业对政府和消费者对政府。最常用的三类是：

消费者在网上购物，通常称为企业对消费者（或 B2C）。

网络上企业之间的交易，通常称为企业对企业（或 B2B）。

公司、政府和其他组织使用互联网技术支持销售和采购活动的交易和业务流程。

随着互联网的快速发展，全球电子商务交易在过去的几十年中有了很大的增长。几乎所有的行业都与电子商务紧密相连。在线市场预计在 2015—2020 年将增长 56%。2017 年，全球零售电子商务销售额达 2.3 万亿美元，同比增长 25%。到 2021 年，电子零售收入预计将增长到 4.88 万亿美元。传统市场预计同期增长率仅为 2%。

在新兴经济体中，中国的电子商务规模每年都在不断扩大。中国拥有 6.68 亿个互联网用户（是美国的两倍），是全球最大的在线市场。2015 年上半年，中国在线销售额达到 2 530 亿美元，占同期中国零售总额的 10%。2016 年达到了 8 990 亿美元。

电子商务市场正以引人注目的速度增长。然而，任何事物都有两面性。一方面，蓬勃发展的电子商务是迄今为止远距离交易最快的方式。它使在家里做生意成为可能，这节省了时间和不必要的手续。这就是电子商务比传统商务更受欢迎的原因。另一方面，也存在许多问题。很难控制虚拟交易，相关法律法规不完善。虚假、欺骗性信息在电子商务中日益增多。没有管理，每次交易都有可能发生损失。所以在电子商务上，每个人都应具有很强的风险意识来保护自己。

第九单元　人工智能

　　人工智能（AI）确切地说是什么？作为开始，我们提供下列定义：
　　AI 是计算机科学的一个分支，它涉及研究和创建显示某种形式智能的计算机系统：学习新概念和新任务的系统、能就我们周围的世界进行推理并得出有用结论的系统、能理解自然语言或理解和领会视觉场景的系统，以及执行需要人的各类智能的其他种类技能的系统。
　　像其他一些对复杂标题的定义一样，对 AI 的理解需要对相关术语的理解，诸如智能、知识、推理、思维、认知、学习和若干计算机相关的术语。尽管我们对这些术语中的很多术语缺少精确的科学定义，但我们能够给出它们的大体定义。当然，本课文的目标之一是给与 AI 相关的所有术语加入特定的意义，包括它们的操作意义。
　　词典把智能定义为获得、理解和应用知识的能力，或者是进行思维和推理的能力。当然，智能不只是这点。它具体体现了有意识地和无意识地通过学习和经验获得的所有知识和技艺：高度精确的视觉和听觉感知；思维；想象；交谈、读、写、驾车、记忆和回忆事实、表达和感受情感的能力，以及更多。
　　智能是这些技艺的集成之和，使我们能回忆起 30 年或更多年未见的面孔，或建造并发送火箭到月球上。这些能力使人类区别于其他的有生命体，并且，如同我们将看到的，这个智能的食粮是知识。
　　我们会一直期望建造显示这些特征的系统吗？此问题的答案是 yes！一些系统早已被开发出来以执行很多种类的智能任务，并且对近期开发的、给人印象更为深刻的系统寄予了很高的期望。
　　现在有一些系统能从例子、从被告知的、从过去相关的经验和通过推理进行学习。有一些系统能解决数学方面的、调度多种多样任务方面的、寻找最佳系统配置方面的、计划军事和商业的复杂策略方面的、诊断医学疾病方面的复杂问题，还有其他一些复杂系统，仅举几个例子。有些系统能"理解"一些自然语言的大部分，有些系统的视觉好得足以"识别"照片上、摄像机和其他传感器拍摄的图像上的物体。有些系统能够以不完备的和不确定的事实进行推理。显然，关于这些开发，自数字计算机问世以来，很多已完成了。
　　尽管有这些印象深刻的成就，我们仍然不能生产具有三岁小孩有的某些基本能力的协调而自主的系统，这些包括识别和记忆一个景象中众多的各种各样的对象、学习新的声言和把它们同对象与概念相关联，以及欣然适应多种多样的新情况的能力。这些都是 AI 研究人员现在面临的挑战，并且它们都不是容易解决的问题。在我们期望能获得比得上三岁小孩的能力之前，在这些能力方面将需要一些重要的突破。

为获得对 AI 更好的理解，知道 AI 不是什么也是有用的。AI 不是研究和创建常规的计算机系统。即使有人可能争辩说所有程序都显示出某种程度的智能，但 AI 程序将超过它，AI 程序表现出的高级智能（在一定程度上）等于或超过了人在完成某项任务中所需要的智能。

AI 不是研究头脑，不是研究肉体，也不是研究语言，如同惯常在心理学、生理学、认知科学或语言学等领域中发现的那样。诚然，在这些领域和 AI 间存在有某种交叠。所有人都在寻求对人的智能和感觉过程有更好的理解。

但是就 AI 而论，目标是开发一些运转计算机的系统，它们真正能执行一些需要高级智能的任务。程序未必打算模仿人的感觉和思维过程。确实，在执行某些不同的任务中它们实际上会超过人的一些能力。重要的一点是这些系统都能有效而高效率地执行智能任务。

最后，看看组成整个 AI 的部分研究领域可更好地理解 AI。这些包括诸如机器人、存储组织、知识表示、存储和恢复、学习法模型、推断技术、常识推理、在推理和决策中不确定性地处理、理解自然语言、模式识别和机器视觉方法、检索和匹配、语音识别和合成，以及各种各样的 AI 工具。

迄今在 AI 方面我们已获得了多少成就？下一个大挑战是什么？这些问题的答案形成本课文中涉及的大部分资料。这里我们仅提到：AI 正进入这样的年代，即实用的商业产品现在已可用了，包括各种各样的机器人设备、识别形状和对象的视觉系统、执行很多困难任务的专家系统（这些系统做得与人类专家一样好或许更好）、帮助调整学生的学习，并监控学生学习进度的智能教育系统、帮助用户建造专门知识库的"智能"编辑器，以及能学习以改进其性能的一些系统。

第十单元　求职信

River 路 102 号
南京 210000
（025）22244556
2018 年 12 月 30 日

迈克·安德森先生
人力资源部主任
ABC 公司
中心街 168 号
南京 210000

尊敬的安德森先生：

我写信来应聘计算机操作员的工作。随函附上 12 月 27 日扬子晚间招聘电脑操作员的广告。我期望成为这个职位的候选人。

我今年 20 岁，毕业于江苏省联合技术学院。我的专业是计算机网络与技术。我主修的课程包括计算机程序设计、AutoCAD、Visual Basic、网络与通信、OA 设备等。在四年的学习期间，我的学业成绩一直名列前茅，两次荣获"优秀学生干部"称号，三次荣获"一流奖学金"。此外，我还拿到了 PETS-3 和计算机二级的证书。值得一提的是，我大二的时候加入了我们学校的计算机图文工作室。我在工作室负责的是编辑图片和文件。在工作室学习了一年之后，我加入了我们学校的网络中心。虽然我只是那里的一名程序员，但是我学到了很多东西，并且我熟悉了用 C#和 ASP. NET 编写网站的过程。根据四年的学习和实践，我认为我有能力胜任这个工作。

我善于探究、善于分析，我喜欢发现事物。我机智、幽默、有着能吸引别人注意力的能力。

如果这些符合您的要求，请准予我面试。

随函附上我的个人简历、近照和证书复印件。

我期待着您的来信。

谨上，

王兵

2018 年 12 月 30 日

附录：
计算机专业英语词汇表

A

a flashing [ˈflæʃɪŋ] cursor [ˈkɜːsə]	闪动的光标
abstraction [əbˈstrækʃ(ə)n]	n. 抽象化
academic [ˌækəˈdemɪk]	adj. 学校的；学院的
academically [ˌækəˈdemɪkəlɪ]	adv. 从学术观点看
access [ˈækses]	n. 调取；存取
accessible [əkˈsesɪb(ə)l]	adj. 易接近的；可理解的；易相处的
accomplish [əˈkʌmplɪʃ]	vt. 完成；实现；达到
advent [ˈædvɜnt]	n. 到来；出现
advertisement [ədˈvɜːtɪsmənt]	n. 广告，宣传；公告；出公告，做广告
allocate [ˈæləkeɪt]	vt. 分配
analog [ˈænəlɒg]	n. 类似物，相似物
analogous [əˈnæləgəs]	adj. 相似的，模拟的
analytical [ˌænəˈlɪtɪkəl]	adj. 分析的，分析法的；善于分析的
Android [ˈændrɔɪd]	n. 基于 Linux 平台的开源手机操作系统
animation [ˌænɪˈmeɪʃ(ə)n]	n. 动画
anti-virus software	杀毒软件
application [ˌæplɪˈkeɪʃ(ə)n]	n. 应用
architect [ˈɑːkɪtekt]	n. 建筑师
architecture [ˈɑːkɪtektʃə]	n. 建筑；架构
arrange [əˈreɪndʒ]	v. 排列；整理
array [əˈreɪ]	n. 展示；陈列；一系列
artificial [ˌɑːtɪˈfɪʃ(ə)l]	adj. 人造的；仿造的
aspect [ˈæspekt]	n. 方面；方向
assemble [əˈsemb(ə)l]	v. 组装
assign [əˈsaɪn]	v. 分配；指定
attempt [əˈtem(p)t]	v. 企图
audio [ˈɔːdɪəʊ]	adj. 音频的
authorized [ˈɔːθəraɪzd] user	授权用户
autonomous [ɔːˈtɒnəməs]	adj. 自治的；自主的；自发的
available [əˈveɪləb(ə)l]	adj. 有空的

| avionics [ˌeɪviˈɒnɪks] | n. 航空电子设备 |
| awareness [əˈweənəs] | n. 意识，认识 |

B

Bios（Basic Input Output System）	基本输入输出系统
booming [ˈbuːmɪŋ]	adj. 兴旺的；繁荣的
boot [buːt]	v. 启动
breakthrough [ˈbreɪkθruː]	n. 突破；突破性进展

C

cabinet [ˈkæbɪnət]	n. 机箱；柜子
camp [kæmp]	n. 露营；营地；阵营
candidate [ˈkændɪdeɪt]	n. 候选人；应试者
capacity [kəˈpæsɪti]	n. 能力；容量
categorize [ˈkætəgəraɪz]	vi. 分类
category [ˈkætəgəri]	n. 种类；类别
click [klɪk]	v. 点击；敲击
Cloud Computing	云计算
cognition [kɒgˈnɪʃ(ə)n]	n. 认识；知识；认识能力
collaborative [kəˈlæbəretɪv]	v. 协作，合作
collapse [kəˈlæps]	v. 崩溃；倒塌；折叠
collective [kəˈlektɪv]	adj. 集体的；共同的；集合的
combination [kɒmbɪˈneɪʃ(ə)n]	n. 组合，联合
commercial [kəˈmɜːʃ(ə)l]	adj. 商业的，营利的
communication [kəˌmjuːnɪˈkeɪʃn]	n. 交流；传达；通信；沟通
company [ˈkʌmp(ə)nɪ]	n. 公司
compatibility [kəmˌpætɪˈbɪlɪti] pack [pæk]	n. 兼容包
compile [kəmˈpaɪl]	v. 编译；编制
compiler [kəmˈpaɪlə]	n. 编译程序
component [kəmˈpəʊnənt]	n. 成分；零件 adj. 组成的
compose [kəmˈpəʊz]	v. 由构成
computer case	计算机机箱
computer programming	计算机编程
concentrate [ˈkɒns(ə)ntreɪt]	vt. 集中，专注于
concept [ˈkɒnsept]	n. 观念，概念
consume [kənˈsjuːm]	v. 消耗；消费；耗尽；毁灭
consumer [kənˈsjuːmə]	n. 消费者
convenient [kənˈviːnɪənt]	adj. 方便的
conventional [kənˈvenʃ(ə)n(ə)l]	adj. 常见的，传统的

converse [kən'vɜːs]	adj. 相反的，逆向的；颠倒的
convert [kən'vɜːt]	v. 转变
coordinate [kəʊ'ɔːdɪneɪt]	vt. 协调
coordinated [kəʊ'ɔːdɪneɪtɪd]	vt. 调整；使调和（coordinate 的过去分词）；调节；整理 adj. 协调的
corruption [kə'rʌpʃ(ə)n]	n. 腐败；贪污；贿赂
cost-effective ['kɔːstə'fektɪv]	adj. 有成本效益的，划算的；合算的
CPU (Central Processing Unit)	n. 中央处理器
craft [krɑːft]	v. 精巧地制作
create [kriː'eɪt]	vt. 创造；创作；创建；创设；设计
customarily ['kʌstəm(ə)rɪlɪ]	adv. 通常，习惯上

D

data ['deɪtə]	n.（datum 的复数）资料，材料；[计] 数据，资料
database ['deɪtəbeɪs]	n. 数据库，资料库
deadline ['dedlaɪn]	n. 最后期限
decade ['dekeɪd]	n. 十年，十年期
deceptive [dɪ'septɪv]	adj. 欺诈的；虚伪的
default [dɪ'fɔːlt]	n. 默认
define [dɪ'faɪn]	vt. 定义
definition [defɪ'nɪʃ(ə)n]	n. 定义
demonstrate ['demənstreɪt]	vt. 证明；展示；论证
desktop ['desktɒp]	n. 台式机 adj. 桌面的
device [dɪ'vaɪs]	n. 设备；装置；终端
diagnose ['daɪəgnəʊz]	vt. 诊断；断定
diagram ['daɪəgræm]	n. 图表；示意图
digital ['dɪdʒɪt(ə)l]	adj. 数字的；数据的；数码的
dimensional [dɪ'menʃənəl]	adj. 空间的
document ['dɒkjʊm(ə)nt]	n. 文档

E

economy [ɪ'kɒnəmɪ]	n. 经济；经济体
edit ['edɪt]	v. 编辑
effective [ɪ'fektɪv]	adj. 有效的；生效的
efficient [ɪ'fɪʃnt]	adj. 有效的；高效率的
elusive [ɪ'l(j)uːsɪv]	adj. 难以捉摸的；逃避的
embedded [ɪm'bedɪd]	adj. 嵌入式的，植入的，内含的
embody [ɪm'bɒdɪ]	vt. 体现，使具体化；具体表达

emerge [ɪˈmɜːdʒ]	vi.	浮现，出现，显现
emergency [ɪˈmɜːdʒ(ə)nsɪ]	n.	紧急情况，突发事件
enclosed [ɪnˈkləuzd]	adj.	被附上的
enhance [ɪnˈhɑːns]	v.	提高；增加
enterprise [ˈentəpraɪz]	n.	企（事）业单位；事业
entity [ˈentɪtɪ]	n.	实体；存在；本质
environment [ɪnˈvaɪrənmənt]	n.	环境
equipment [ɪˈkwɪpm(ə)nt]	n.	设备，装备；器材
executable [ɪgˈzekjʊtəb(ə)l]	adj.	(可)执行的；实行的
execute [ˈeksɪkjuːt]	v.	执行；完成；履行
exist [ɪgˈzɪst]	v.	存在
expand [ɪkˈspænd]	vi.	扩张；发展；膨胀
expert [ˈekspɜːt]	n.	专家；能手；权威；行家；高手

F

facilitate [fəˈsɪlɪteɪt]	v.	促进；帮助
facilitation [fəsɪlɪˈteɪʃn]	n.	简易化
facility [fəˈsɪlətɪ]	n.	工具；设备
feat [fiːt]	n.	技艺表演
filename [ˈfaɪlnem] extension [ɪkˈstenʃ(ə)n]	n.	文件扩展名
floppy [ˈflɒpɪ] disk	n.	软盘驱动器
fluency [ˈfluːənsɪ]	n.	流畅；流畅性
folder [ˈfəuldə]	n.	文件夹
formality [fɔːˈmælətɪ]	n.	礼节；仪式；正式手续
format [ˈfɔːmæt]	n.	版本，形式
former [ˈfɔːmə]	adj.	前者的
fragile [ˈfrædʒaɪl]	adj.	易碎的
function [ˈfʌŋ(k)ʃ(ə)n]	n.	功能
fund [fʌnd]	n.	资金；基金

G

generate [ˈdʒenəreɪt]	v.	产生
global [ˈgləub(ə)l]	adj.	全球的；全世界的
grant [grɑːnt]	vt.	授予；允许；承认 vi. 同意
graphics [ˈgræfɪks]	n.	图表

H

haptic [ˈhæptɪk]	adj.	触觉的
hard disk		硬盘

hardware [ˈhɑːdweə]	n. 计算机硬件
Homo sapiens	智人（现代人的学名）；人类
hypertext [ˈhaɪpətekst]	n. 超文本

I

identical [aɪˈdentɪk(ə)l]	adj. 完全相同的；同一的
image [ˈɪmɪdʒ]	n. 图像
impart [ɪmˈpɑːt]	vt. 给予（尤指抽象事物），传授；告知，透露
infect [ɪnˈfekt]	v. 感染；传染；散布病毒；侵染
inference [ˈɪnf(ə)r(ə)ns]	n. 推理；推论；推断
infrastructure [ˈɪnfrəstrʌktʃə]	n. 基础设施；基础结构；基础建设
innovation [ˌɪnəˈveɪʃn]	n. 革新，创新
inquisitive [ɪnˈkwɪzɪtɪv]	adj. 好奇的；爱打听的；求知欲强的
insert [ɪnˈsɜːt]	v. 插入
install [ɪnˈstɔːl]	v. 安装
institute [ˈɪnstɪtjuːt]	n. 协会；学会；学院
instruction [ɪnˈstrʌkʃ(ə)n]	n. 指令
integrate [ˈɪntɪgreɪt]	v. 使一体化；使整合
integrated [ˈɪntɪgreɪtɪd]	adj. 集成的；综合的；完整的；互相协调的
intelligence [ɪnˈtelɪdʒ(ə)ns]	n. 智力
interactivity [ˌɪntərˈæktɪvɪtɪ]	n. 互动
interchange [ˈɪntətʃendʒ]	n. 互换
interconnect [ˌɪntəkəˈnekt]	vi. 互相连接；互相联系
interface [ˈɪntəfeɪs]	n. 接口，交界面
intervening [ˌɪntəˈviːnɪŋ]	adj. 中介的，介于其间的

K

keyboard [ˈkiːbɔːd]	n. 键盘

L

laptop [ˈlæptɒp]	n. 笔记本计算机
latter [ˈlætə]	adj. 后者的
linear [ˈlɪnɪə]	adj. 线性的
linguistics [lɪŋˈgwɪstɪks]	n. 语言学

M

main [meɪn] duty [ˈdjuːtɪ]	主要工作职责
mainframe [ˈmeɪnfreɪm]	n. 主机；大型机
malicious [məˈlɪʃəs]	adj. 生殖的；再生产的；复制的

mechanism ['mek(ə)nɪz(ə)m]	n. 机能；机制；结构
metaphor ['metəfə]	n. 隐喻
methodology [meθə'dɒlədʒɪ]	n. 方法论；方法学
military ['mɪlɪt(ə)rɪ]	adj. 军事的；军人的；适于战争的
modem ['məʊdem]	n. 调制解调器
monitor ['mɒnɪtə(r)]	n. 显示器
motherboard ['mʌðəbɔːd]	n. 主板，主机板
mouse [maʊs]	n. 鼠标
multimedia ['mʌltɪmiːdɪə]	n. 多媒体
multiple ['mʌltɪpl]	adj. 多重的；复杂的；多功能的
must-have	n. 必需品

N

navigation [nævɪ'geɪʃ(ə)n]	n. 航海；航空；航行
navigational [ˌnævɪ'geɪʃnəl]	adj. 航海的，航行用的
nonlinear [nɒn'lɪnɪə]	adj. 非线性的
noticeable ['nəʊtɪsəb(ə)l]	adj. 显而易见的；显著的；值得注意的

O

objective [ˌəb'dʒektɪv]	n. 方法论；方法学
object-oriented ['ɒbdʒekt ɔːrɪentɪd]	adj. 面向对象的；对象趋向的
official [ə'fɪʃ(ə)l]	adj. 官方的；正式的；官方认可的
online [ɒn'laɪn]	adj. （计算机）联机的；（与计算机）联线的
operator ['ɒpəreɪtə]	n. 经营者；操作员
optimal ['ɒptɪm(ə)l]	adj. 最佳的；最理想的
organization [ˌɔːgənaɪ'zeɪʃn]	组织；团体；机构
orient ['ɔːrɪent]	n. 东方
OS (operating system)	n. 操作系统
overlap [əʊvə'læp]	n. 重叠；重复

P

package ['pækɪdʒ]	n. 包
participate [pɑː'tɪsɪpeɪt]	vi. 参与，参加
part-time job	兼职工作
password ['pɑːswɜːd]	n. 密码，口令
perceive [pə'siːv]	vt. 察觉，感觉；理解；认知
perception [pə'sepʃ(ə)n]	n. 知觉；[生理] 感觉；看法；洞察力；获取
performance [pə'fɔːm(ə)ns]	n. 性能
perspective [pə'spektɪv]	n. 观点；远景

physiology [ˌfɪzɪˈɒlədʒɪ]	n. 生理学
pirate [ˈpaɪərət] software	盗版软件
platform [ˈplætfɔːm]	n. 平台
portable [ˈpɔːtəb(ə)l]	adj. 手提的；便携式的
precise [prɪˈsaɪs]	adj. 精确的；明确的；严格的
preferable [ˈpref(ə)rəb(ə)l]	adj. 更好的，更可取的，更合意的
presentation [prez(ə)nˈteɪʃ(ə)n]	n. 显示，表演
preventive [prɪˈventɪv]	v. 预防；防止；预防措施
previous [ˈpriːvɪəs]	adj. 以前的，早先的
primary [ˈpraɪm(ə)rɪ]	adj. 主要的；初级的；基本的
prior [ˈpraɪə]	adj. 在前的
process [ˈprəʊses]	v. 处理
program [ˈprəʊɡræm]	n. 节目；程序 v. 编制程序
project [prəˈdʒekt]	v. 放映，投影
property [ˈprɒpətɪ]	n. 特性；属性；财产
proprietary [prəˈpraɪət(ə)rɪ]	adj. 所有的；专利的；私人拥有的
protocol [ˈprəʊtəkɒl]	n.（信息交换）协议
psychology [saɪˈkɒlədʒɪ]	n. 心理学
publish [ˈpʌblɪʃ]	v. 出版；印刷
purchase [ˈpɜːtʃəs]	vt. 购买；赢得
purpose [ˈpɜːpəs]	n. 目的

R

random [ˈrændəm]	adj. 随便的；任意的
reasoning [ˈriːz(ə)nɪŋ]	n. 推理；论证；评理 adj. 推理的
receipt [rɪˈsiːt]	n. 收据；收入
refine [rɪˈfaɪn]	vt. 精炼，提纯；改善；使……文雅
regulation [reɡjʊˈleɪʃ(ə)n]	n. 管理；规则
relevant [ˈreləvənt]	adj. 相关的
remote [rɪˈməʊt]	adj. 远程的；遥远的
replicate [ˈreplɪkeɪt]	v. 复制；复写；重复；反复
reply [rɪˈplaɪ]	v. 答复；回应 n. 回答；答复
resolution [rezəˈluːʃ(ə)n]	n. 分辨率
restriction [rɪˈstrɪkʃ(ə)n]	n. 限制；限定；拘束；束缚；管制
resume [rɪˈzjuːm]	n. 摘要；[管理] 履历，简历
retail [ˈriːteɪl]	n. 零售
revenue [ˈrevənjuː]	n. 税收收入；财政收入；收益
roughly [ˈrʌfli]	adv. 大约，大致

S

safe mode	安全模式
scanner [ˈskænə]	n. 扫描仪
scope [skəup]	n. 处理，研究事务的范围
screw [skruː] up	弄糟
secondary [ˈsek(ə)nd(ə)rɪ]	adj. 次要的；第二的
self-paced [selfˈpeɪst]	adj. 自定进程的，自定步调的
sensor [ˈsensə]	n. 传感器
series [ˈsɪəriːz]	n. 系列；连续
shutdown [ˈʃʌtdaun]	n. 关机
simulation [ˌsɪmjuˈleɪʃn]	n. 模仿；模拟
software [ˈsɒf(t)weə]	n. 软件
sophistication [səˌfɪstɪˈkeɪʃn]	n. 复杂程度
sophomore [ˈsɒfəmɔː]	n. 大学（或中学）二年级学生
source [sɔːs]	n. 来源；出处
speak at length about	详细讲述
spreadsheet [spredʃiːt]	n. 表格
standard [ˈstændəd]	adj. 普通的；正常的；通常的；标准的
storage [ˈstɔːrɪdʒ]	n. 储藏；仓库；[计] 存储器
strategy [ˈstrætədʒɪ]	n. 策略；战略；战略学
stream [striːm]	n. 流
subscription [səbˈskrɪpʃ(ə)n]	n. 认购；预订；订阅
supplier [səˈplaɪə]	n. 供应商
surf [sɜːf]	vi. 滑浪；（互联网上）冲浪；漫游
synthesis [ˈsɪnθɪsɪs]	n. 综合，[化学]合成；综合体
system [ˈsɪstəm]	n. 体系，系统；制度
System Restore	n. 系统恢复

T

tablet [ˈtæblɪt]	n. 平板计算机
tact [tækt]	n. 机智，机敏；老练
tamper [ˈtæmpə]	v. 篡改；窜改
TCP/IP	传输控制协议/因特网互联协议
teamwork spirit	团队合作精神
textile [ˈtekstaɪl]	n. 纺织品
theft [θeft]	v. 偷盗；偷窃；被盗
threat [θret]	v. 威胁；恐吓；凶兆
transaction [trænˈzækʃ(ə)n]	n. 交易，事务

transcend [træn'send]	v. 超出，超过
transfer [træns'fɜː]	传递；转移
transmit [trænz'mɪt]	v. 传送，传播
trillion ['trɪljən]	n. (数) 万亿

U

underlying [ˌʌndə'laɪɪŋ]	adj. 潜在的；含蓄的；基础的
uninstall [ˌʌnɪn'stɔːl]	v. 卸载 n. 解除安装
USB flash disk	U 盘

V

variety [və'raɪəti]	n. 多样；种类
via ['vaɪə]	prep. 经由，通过
video conferencing	电视会议
virtual ['vɜːtʃuəl]	adj. 事实上的，实际上的；(计) 虚拟的
virtually ['vɜːtʃuəli]	adv. 事实上地，实际上地
virus ['vaɪrəs]	n. 病毒
virus ['vaɪrəs] scanner ['skænə]	病毒扫描软件

W

warehouse ['weəhaʊs]	n. 仓库

附录：

计算机专业英语专业术语词汇表

A

a multi-user system	多用户系统
a network file system	网络文件系统
application program	应用程序
application software	应用软件
application suite	成套应用软件
artificial intelligence	人工智能
AutoCAD	计算机绘图
automated data collection system	自动数据收集系统
automated teller machines（ATM）	自动柜员机；ATM 机

B

business-to-business（B2B）	企业对企业
business-to-consumer（B2C）	企业对消费者
business-to-government（B2G）或（B2A）	企业对政府

C

commercial software package	商业软件包
commonsense reasoning	常识推理
computer aided design（CAD）	计算机辅助设计
computer aided engineering（CAE）	计算机辅助工程
computer application	计算机应用
computer program designing	计算机程序设计
computer security	计算机安全
computer technology	计算机技术
computer virus	计算机病毒
consumer to government（C2G）或（C2A）	消费者对政府
consumer-to-consumer（C2C）	消费者对消费者
CPU time	CPU 时序
CPU（Central Processing Unit）	中央处理器

D

database management system（DBMS）	数据库管理系统
desktop computer	台式计算机
Desktop OS	桌面操作系统
desktop publishing software	桌面印刷软件
digital camera	数码相机
digital electronic media	数字式电子媒介技术
digital online multimedia	数字式网上多媒体
dishonest adware	不可信任的恶意广告
display screen	显示屏
distributed environment	分布环境

E

e-commerce	电子商务
educational software	教育软件
electronic data interchange（EDI）	电子数据交换
electronic funds transfer（EFT）	电子资金转移
embedded Linux	嵌入式 Linux
Embedded OS	嵌入式操作系统
emerging economy	新兴经济体
enterprise software	企业软件
e-retail revenue	电子零售收入

F

fax machine	传真机
file name text	文件名文本框
file storage space	文件储存空间

G

graphics software	图形软件

H

hardware architecture	硬件架构
hardware resource	硬件资源
household appliance	家用电器

I

inference technique	推理技术
information security	信息安全

input device	输入设备
input/output (I/O) device	输入输出设备
integrated software	集成软件
intended user	预期用户
interact with	与……相互作用
Internet marketing	互联网营销
inventory management system	库存管理系统

L

Linux distribution	Linux 发行版

M

main memory	主存
mainframe computer	大型计算机
media development	媒体开发软件
memory organization	存储组织
memory space	储存空间
Microsoft Office	微软办公软件
mobile application software	移动应用软件
mobile commerce	移动商务
multimedia presentation	多媒体演示

N

network and communications	网络与通信

O

OA equipment	办公自动化设备
online shopping sales	在线销售额
online transaction processing	在线交易处理
operating system	操作系统
output device	输出设备

P

preventive measure	预防措施
print layout view	打印预览
product engineering software	产品工程软件

R

RAM (Random Access Memory)	随机存取存储器
related function	相关函数

resource allocator	资源分配器
retail e-commerce sales	零售电子商务销售额
robotic	机器人
ROM（Read-only Memory）	只读存储器

S

search engine	搜索引擎
secondary/auxiliary storage	辅助存储器
sensor	传感器
Server OS	服务器操作系统
simulation software	仿真软件
still image	静态图像
supply chain management	供应链管理
system software	系统软件

T

tablet computer	平板计算机
target diskette	目标磁盘
telephone banking	电话银行
the collective processes	系列过程
the reproductive ability	复制能力
total Chinese consumer retail sales	中国零售总额

U

underlying capability	潜在能力
Unix-like OS	类 Unix 操作系统
untrustworthy individual	不可信任的个体
USB（Universal Serial Bus）disk	U 盘
user interface	用户界面

V

Visual Basic，VB	一种通用的基于对象的程序设计语言

W

watch out	提防；小心
word processor	文字处理器
World Wide Web	万维网

— 115 —

练习答案

Unit 2 Computer Hardware

Warm-up

Match the pictures with the words and phrases, then write the correct letter in the brackets to each.

(B) 1. modem (D) 2. fax machine
(C) 3. notebook computer (A) 4. desktop computer

Section 4: Exercises

Ⅰ. The following pictures show some hardware of computer. Please label the pictures with the given words and phrases.

mouse __G__ CPU __J__ motherboard __A__
printer __C__ keyboard __B__ graphic card __D__
memory __H__ hard disk __I__ monitor __F__ iPod __E__

Ⅱ. Fill in the blanks with some of the above mentioned words and phrases.

Computer hardware consists of:

1. Input device: <u>Mouse/Keyboard</u> is one of the common input devices.

2. Processor unit: It is divided into <u>CPU</u> and main memory.

3. Output device: <u>Monitor</u> and <u>printer</u> are the two most commonly used output devices.

4. Auxiliary storage unit: The common auxiliary storage devices are <u>iPod</u> and a hard disk drive.

Ⅲ. Put the verbs in the passage into correct tenses.

Some people <u>say</u> (say) that we live in the age of computers, but it <u>is</u> (be) also correctly described as the atomic age or the space age. Today, a journey from London to Cairo <u>takes</u> (take) hours. Only a hundred years ago it <u>took</u> (take) weeks. Today, men <u>think</u> (think) seriously of going to Mars. Fifty years ago they only <u>dreamed</u> (dream) about it. Today, we <u>produce</u> (produce) energy by splitting the atom. A century ago, no one <u>believed</u> (believe) it could be <u>split</u> (split). Technology <u>advances</u> (advance) so quickly that cars and televisions <u>are</u> (be) out of date only a few years after they <u>were</u> (be) made.

Ⅳ. Ability to explore.

1. D   2. A、B、C、E

Ⅴ. Fill in the table below by giving the corresponding translation with the help of dictionaries or Internet.

English	Chinese
telecommunications industry	电信产业
bandwidth	带宽
information age	信息化时代
digital technology	数码技术
electronics technology	电子科技
business licence	营业执照
palm-sized computer	掌上电脑
audiovisual products	视听产品
download free and shared software	下载免费共享软件
analog signal	模拟信号

Section 5：Furthering Reading

1. The first mouse had two wheels set at a 90-degree angle to each other to keep track of the movement.

2. It was invented around 1980.

3. It received wide use only after IBM introduced their branded laser printer known as IBM 3800 in 1976.

4. No, today their use has become history.

5. DO NOT POWER IT DOWN.

Unit 3 Operating System

Warm up

1. 视窗系统架构   2. 视窗系统可执行文件

3. 应用程序库   4. 视窗系统动态链接库

5. 图形设备接口   6. 内核接口

7. 应用程序   8. 实用程序

9. 编译程序   10. 命令外壳程序

Section 2：Reading

Pre-reading Activity

2. 3. 4. 6. 7.

Tell whether the following statements are true (T) or false (F).

1. F   2. T   3. F   4. T   5. F

Section 4: Exercises

Ⅰ. Fill the blanks with the words or phrases of the text.
1. components   2. resources   3. interface   4. allocator
5. desktop OS, server OS, embedded OS

Ⅱ. Fill the blanks with the following words or phrases. Change the form if necessary.
1. refer to   2. provide   3. primary   4. manage
5. a wide variety of   6. in an efficient way   7. acted as   8. was divided into

Ⅲ. Translate the following sentences into Chinese.
1. 计算机系统大致可以分为四个部分：硬件、操作系统、应用软件和用户。
2. 可能有许多不同的用户试图解决不同的问题，所以可能有很多不同的应用程序。
3. 操作系统(OS)是管理计算机硬件和软件资源并为计算机程序提供公共服务的系统软件。
4. 计算机系统有许多解决问题可能需要的资源：CPU 时序、内存空间、文件存储、输入/输出（I/O）设备，等等。
5. 嵌入式系统广泛应用在生活的各个方面，涵盖范围从便携设备到大型固定设施，如数码相机、手机、平板计算机、家用电器、医疗设备、交通灯、航空电子设备和工厂控制设备等。

Ⅳ. Ability to explore.
B

Ⅴ. Fill in the table below by giving the corresponding translation with the help of dictionaries or Internet.

English	Chinese
software package	软件包
document management	文档管理
fragmentation	碎片整理
image capture	影像采集
text formatting	文本格式化
interactive	交互式处理
hardware compression	硬件压缩
server system	服务器系统

Section 5: Furthering Reading
1. B   2. A   3. C

Unit 4   Application Software

Warm-up

Ⅰ. All are general office automatic devices.

Ⅱ. A. D.

Pre-reading Activity

1. E 2. F 3. H 4. B 5. G 6. D 7. C 8. A

Mark the following sentences with true or false according to the passage.

1. T 2. F 3. F 4. T 5. F 6. F 7. T 8. F

Section 4: Exercises

Ⅰ. Why are operating systems considered inseparable from the hardware? Which of the following reasons is not true?

B

Ⅱ. Complete the sentences below with the correct verbs.

1. shows   2. enter   3. Save   4. insert   5. exists
6. upgrade   7. compile   8. interact   9. relate   10. integrate

Ⅲ. Fill in the blanks according to the passage.

1. software
2. system software; application software
3. system
4. application
5. multiple applications, functions, features and user interfaces
6. educators, students
7. mobile
8. hardware, software

Ⅳ. Reading comprehension.

1. F 2. T 3. F 4. T 5. T

What you want to do	What kind of software you need	Example
Create text (letters, term papers, etc.)	A	Word
Set up a worksheet	D	Excel
Build a presentation	E	PowerPoint
Build Web sites	B	Frontpage
Communicate with other people	C	Outlook
Deal with a large amount of data, calculations	F	Access

Section 5: Furthering Reading

1. C 2. D 3. D 4. D

Unit 5   Multimedia

Warm-up

1. B 2. F 3. C 4. A 5. D 6. E

Section 2: Reading

1. B 2. C 3. C 4. C 5. A 6. A

— 119 —

Section 4: Exercises

Ⅰ. Tell whether the following statements are True (T) or False (F).

(　) 1. F
(　) 2. F
(　) 3. T
(　) 4. T
(　) 5. F
(　) 6. F
(　) 7. T
(　) 8. F

Ⅱ. Complete the following sentences according to the text.

1. text, audio, still images, animation, video, interactivity content forms
2. linear, non-linear
3. live, recorded
4. a navigation system, an interaction with the presenter or performer.
5. a live or recorded multimedia presentation.
6. enhance levels of interactivity

Ⅲ. Ability to explore.

(　) 1. C
(　) 2. D

Ⅳ. Fill in the table below by giving the corresponding translation with the help of dictionaries or Internet.

English	Chinese
Audio/Video Interleave (AVI)	音频/视频交错格式
API (Application Programming Interface)	一种程序接口
Joint Photographic Experts Group (JPEG)	联合图像专家组
Moving Picture Expert Group	运动图像专家组
National Television Standards Committee (NTSC)	美国全国电视标准委员会
Phase Alternating Line, PAL	逐行倒相制（电视）

Section 5: Furthering Reading

Ⅰ. 1. 图片处理　2. 图像编辑　3. 矢量图形　4. 圆角矩形　5. 格式控制

Ⅱ. 1. T　2. F

Unit 6　Computer Networks

Warm-up

Omitted.

Section 4: Exercises

Ⅰ. The pictures below are about computer. Please label the pictures and complete the following sentences with the given words and phrases.

1. E-mail   B   2. FTP   D
3. WWW   C   4. Telnet   A

1.   D   is a method of transferring files from one computer to another over the Internet, even if each computer has a different operating system or storage format.

2. The   C   , which Hypertext Transfer Protocol (HTTP) works with, is the fastest growing and most widely-used part of the Internet.

3.   A   allows an Internet user to connect to a distance computer and use that computer as if he or she were using directly.

4. The most widely used tool on the Internet is   B   . It enables you to send messages to other areas, no matter how far between individuals.

Ⅱ. Read the text and decide whether the following statements are True or False.

(T) 1. The Internet is a huge network, connecting computers to computers across the world.
(F) 2. If you don't know the address, you could never get to the website you want.
(T) 3. Search engine is a useful tool when surfing the Internet.
(F) 4. Only big companies and governments are allowed to create a website.
(F) 5. The Internet is powerful and everything in it is true.

Ⅲ. Choose the best answer according to the passage.

1. A   2. C   3. D   4. C   5. D

Ⅳ. Ability to explore

D

Ⅴ. Fill in the table below by giving the corresponding translation with the help of dictionaries or Internet.

English	Chinese
DNS (Domain Name Server)	域名服务器
FTP (File Transfer Protocol)	文件传送协议
ISDN (Integrated Services Digital Network)	综合业务数字网
LAN (Local-area Network)	局域网
POP (Post Office Protocol)	邮局协议
PPP (Peer-Peer Protocol)	对等协议
SLIP (Serial Line Interface Protocol)	串行线接口协议
SMTP (Simple Message Transfer Protocol)	简单邮件传送协议
TCP/IP (Transfer Control Protocol/Internet Protocol)	传送控制协议
URL (Uniform Resources Locator)	统一资源定位器

Section 5：Furthering Reading
1. e 2. c 3. d 4. a 5. b

Unit 7 Computer Security

Warm-up
1. f 2. b 3. a 4. d 5. c 6. e

Section 2：Reading

Pre-reading Activity
Ⅰ. Below are some possible symptoms of a virus. Translate them into Chinese.
1. 系统速度变慢
2. 系统在启动一分钟后反复重启
3. 你的密码未经提示却改变了
4. 一些文件没有删除却丢失了
5. 其他人知道了你的密码或其他私人信息
6. 某一程序运行后消失了

Ⅱ. Match the following viruses with their Chinese meanings.
1. c 2. f 3. d 4. a 5. b 6. e

Mark the following sentences with true or false according to the passage.
1. T 2. T 3. F 4. T 5. T

Section 4：Exercises
Ⅰ. Match the items listed in the following two columns.
1. c 2. b 3. d 4. a 5. e 6. f 7. g 8. h

Ⅱ. Which are the potential possibilities if your computer breaks down?
B

Ⅲ. Translation.
1. as Information Security as applied to computers and networks
2. imposes restrictions on that program's behavior
3. even those that do not have the reproductive ability
4. that can detect and eliminate known viruses
5. replicate by attaching a copy of themselves to another program
6. to be regularly updated in order to recognize the latest threats

Ⅳ. Reading comprehension.
1. B 2. D 3. B 4. A

Section 5：Furthering Reading
1. A 2. D 3. C 4. B

Unit 8　E-commerce

Warm-up

The shopping sites are Alibaba, dangdang.com, ebay, amazon, 1号店, Taobao.com

Section 2：Reading

Pre-reading Activity

Ⅰ. Try to match the expressions with their Chinese meanings.

1. b　2. e　3. d　4. a　5. c

Ⅱ. Have you heard of the following terms of e-commerce? Try to match the activities with their definitions and see if you are right after reading the passage.

1. c　2. b　3. a

Answer these questions for comprehension.

1. E-commerce is short for "electronic commerce."

2. E-commerce is the activity of buying or selling of products on online services or over the Internet.

3. It relies on a variety of technologies such as mobile commerce, electronic funds transfer, supply chain management, Internet marketing, online transaction processing, electronic data interchange (EDI), inventory management systems, and automated data collection systems.

4. The five general categories of e-commerce are business-to-consumer, business-to-business, consumer-to-consumer, business-to-government and consumer to government.

5. In 2017, the retail e-commerce sales worldwide amounted to 2.3 trillion US dollars

Section 4：Exercises

Ⅰ. Fill in the following blanks according to the passage.

1. online serves, Internet　　　2. types of entities

3. Business to business　　　　4. Consumer

5. Internet　　　　　　　　　　6. e-commerce presence in China

7. across far distance　　　　　8. losses

Ⅱ. Complete the following sentences.

1. is short for electronic funds transfer.

2. the growth of credit cards, automated teller machines (ATM) and telephone banking

3. closely connected with e-commerce.

4. retail e-commerce sales worldwide

5. the world's biggest online market.

6. at noticeable rates.

Ⅲ. Translate the following sentences into Chinese.

1. 随着互联网的引入，公司无论规模大小，都可以通过电子方式进行交流，而且成本低廉。

— 123 —

2. 吸引新客户比留住现有客户要多花十倍的精力和金钱。
3. 网上购物在中国越来越受欢迎。
4. Ebay 是世界上最大的互联网拍卖网站。
5. 从客户的角度来看，运输成本是一个主要问题。

Ⅳ. Fill in the blanks with the correct words.

1. gain   2. reduce   3. lower   4. improve   5. client
6. reap   7. products   8. convenience   9. information   10. competitive

Ⅴ. Ability to explore.

C. E. D. A. B

Ⅵ. Fill in the table below by giving the corresponding translation with the help of dictionaries or Internet.

English	Chinese
logistics service	物流服务
business model	商业模式
small and medium-sized firms	中小型企业
bank account	银行账户
cancel the deal	取消交易
raw material	原材料
second-hand market	二手市场
lower the cost	降低成本
shipping cost	货运成本
the bottom price	最低价
electronic payment	电子支付
auction website	拍卖网站

Section 5：Furthering Reading

1. online, bookstore   2. Choice, low
3. customer-centered   4. Monitors, adjust, advantage

Unit 9 New Technologies

Warm-up

Omitted.

Section 2：Reading

Choose the best answer according to the text.

1. A   2. C   3. D   4. D   5. B

Section 4: Exercises

Ⅰ. Tell whether the following statements are True (T) or False (F).

1. T
2. F
3. F
4. T
5. F
6. T
7. F
8. F
9. T
10. T

Ⅱ. Complete the following sentences according to the text.

1. learn new concepts and tasks,

can reason and draw useful conclusions about the world around us,

can understand a natural language or perceive and comprehend a visual scene,

perform other types of feats that require human types of intelligence

2. intelligence, knowledge, reasoning, thought, cognition, learning, computer-related terms

3. acquire, understand, apply, thought, reason

4. mind, body, languages, psychology, physiology, cognitive science, linguistics

5. senses, thought

Ⅲ. Ability to explore.

1. D
2. 略

Ⅳ. Fill in the table below by giving the corresponding translation with the help of dictionaries or Internet.

English	Chinese
Machine Learning	机器学习
Natural Language Processing	自然语言处理
Affective Computing	情感计算
Robotics	机器人学
Machine Perception	机器感知
Machine Recognition	机器认知
Knowledge Representation	知识表示

Section 5: Furthering Reading

1. Realcine.
2. Engineers can enter the design of a neighborhood into a computer, and use VR to "walk"

around the neighborhood.
3. The technology behind this product is virtual reality.
4. They wear special VR Headsets and gloves.
5. They could be trained using Realcine without the risk of being sent into a burning building.

Unit 10　Job Application

Warm-up
1. B　2. C　3. A　4. D

Section 2：Reading
Ⅰ. Read the passage and answer the following questions.
1. Computer operator.
2. He saw the post in Yangtse Evening on December 27.
3. He graduated from Jiangsu Union Technical Institute.
4. Yes, he did.

Section 4：Exercises
Ⅰ. Complete the following form with the information you get from the passage.
Name：Wang Bing　　　　Address：102 River Road, Nanjing

Item	Information in an application letter
Education background	1. Graduate from：Jiangsu Union Technical Institute. 2. Standard tests：PETS – 3 and Computer Stage 2.
Relevant courses	Computer Program Designing, Auto-CAD, Visual Basic, Network and Communications, OA Equipment.
Work experience	Part-time jobs in school： 1. He joined the Computer Image & Word processing Studio in our school. 2. He joined the Network Center in our school.
Relevant skills	He has known the process of how to program a website with C# and ASP. NET.
Awards and achievements	He has won the title of "Excellent Student Leader" twice and first-class scholarship three times.

Ⅱ. Put the verbs in the passage into correct tenses：

After university you will start hunting (hunt) for a job, because seeking (seek) a job can be very difficult nowadays. If you want to apply (apply) for a job advertising in a newspaper or magazine, first you must write an application (apply) letter stating that you are interested in it. You then need to fill in an application (apply) form and attach your CV. The CV, similar to a resume (resume) in American English, is the paper stating your working experience or education. If the company needs a reference, you have to ask your boss or professor to be a referee (refer).

Ⅲ. Ability to explore.

Part 1.

1. F 2. A, B, H, J 3. I 4. C 5. D 6. E, G

Part 2

F→B→J→A→H→I→C→G→D→E

Ⅳ. In order to apply for a job, you need to fill out the CV. Fill in the form by yourself.

Omitted

Ⅴ. Fill in the table below by giving the corresponding translation with the help of dictionaries or Internet.

English	Chinese
Application letter	求职信
Interview	面试
CV	简历
Referee	推荐人
Courses taken	所学课程
Required courses	必修课程
Specialized courses	专业课程
"Three Goods" students	三好学生
Major in	主修
Bachelor degree	学士学位

Section 5: Furthering Reading

1. System Engineer

2. Nanjing University of Technology, Computer Science

3. He worked as a student assistance in computer room and a senior system engineer at Beyond Co.

4. Java/J++, C/C++, HTML, X-Window